BIRDS OF SOUTHERN FLORIDA

including THE EVERGLADES, THE KEYS, SANIBEL AND CAPTIVA ISLANDS

G. MICHAEL FLIEG
and ALLAN SANDER

RALPH
CURTIS
BOOKS

First published in 2001 by
New Holland Publishers (UK) Ltd
London • Cape Town • Sydney • Auckland

RALPH CURTIS PUBLISHING, INC.
P.O. Box 349
Sanibel Island, Fl 33957
U.S.A.

ISBN 0 88359 053 0

Publishing Manager: Jo Hemmings
Project Editor: Mike Unwin
Copy Editor: Guy Kirwan
Assistant Editor: Camilla MacWhannell
Design and cartography: D & N Publishing, Baydon, Wiltshire

Reproduction by Modern Age Repro House Limited, Hong Kong
Printed and bound in Malaysia by Times Offset (M) Sdn Bhd

10 9 8 7 6 5 4 3 2 1

Front cover photograph: Roseate Spoonbill (Allan Sander)
Title page photograph: Florida Scrub-jay (Brad Martin)

Acknowledgments
The author wishes to thank Wally George of Fort Lauderdale, my good friend and mentor, who kindly provided current range and distribution data for Southern Florida birds.

Allan would like to thank the contributing photographers for use of their images as well as the following individuals for their support, research, and assistance in the field: Walter George, Mike Lubbock of Sylvan Heights Waterfowl, Travis MacClendon and Audubon's Center for Birds of Prey.

Photographs in this book were taken by Allan Sander except where other photographers' names are indicated next to the photographs.

Contents

Introduction

A Photographic Guide to Birds of Southern Florida illustrates over 250 species that occur in the region either as residents or migrants. Many exotic species are also covered, as this part of the USA has hosted a wide range of deliberate and accidental introductions. Some species, such as Cattle Egret and Shiny Cowbird, have invaded southern Florida naturally, while others, such as Greater Flamingo, Florida Scrub-jay and Red-cockaded Woodpecker, have declined in recent decades. Still others, like the Bahama Mockingbird and Key West Quail-Dove, are casual wanderers from the West Indies. Spectacular waterbirds, including the Anhinga, Roseate Spoonbill, Limpkin, Swallow-tailed Kite, and a variety of herons are abundant, and can often be observed from the road. Many of these are included in this guide. Almost all of the photos have been taken in the field in the birds' natural habitats. The small size of this book enables the user to carry it in a pocket for instant identification.

There have been over 400 species observed in southern Florida. Approximately 350 are native, naturalized and self-supporting through introductions, or reach the region on a regular basis. This guide covers those species most likely to be encountered by birdwatchers. Nearly all of the families found in the area are included. Also depicted are some of the rarer, more interesting and spectacular species.

Careful study of the photographs included in this guide should enable you to identify species' as they are encountered in the field. This work is not a complete guide, and a single photo may be inadequate for certain identification. Behavior, songs and views of a bird from various angles are also important. Notes and sketches will prove invaluable when consulting another reference book, at a later time, to make an identification. A single characteristic may be sufficient to identify some species, but most identifications require use of a combination of features. More specialized books may be required to identify every bird. Both the third edition of the *National Geographic Field Guide to the Birds of North America* and *The North American Bird Guide* contain much additional information on southern Florida birds and are highly recommended, while *A Birder's Guide to Florida*, by Bill Pranty, will assist in locating the best birding sites. See the section on further reading.

Some birdwatching can be done without equipment. Many birds come to feeders, picnic tables and lawns, and are relatively tame. A pair of binoculars is the most important tool for any birder, and their cost varies widely. Research the products on the market carefully for the one that meets your needs and budget.

In practice, the most valuable assistance to a novice is to go into the field with other, more experienced, birdwatchers, and learn some of the tactics that one must employ to be successful in birding. The expertise of experienced birders cannot be over-emphasized.

Many people begin to observe birds while in Florida, as the many large, easily observed birds, spectacular scenery and large expanses of habitat permit one to fully appreciate the entire ecology of the area. Southern Florida, as defined in this book, includes all of the area south of state route 70, broadly from a line drawn from Fort Pierce to Sarasota, and also includes the keys and the Dry Tortugas.

Many birders make a special effort to visit the Miami/Fort Lauderdale area to add the established exotics to their North American life lists. These include several species of parrot, three mynahs, an oriole, and a bulbul, as well as a varied list of additional species.

How to use this book

This book has been designed for clarity and with ease of use in mind. On page 6 is a key to symbols used on each page of the main species descriptions. These symbols will guide the reader to a family or group of families to which each bird belongs. Each such symbol appears on the first full page bearing descriptions of that group. Photographs show the commonly seen plumage. Where there are two or more photographs they are generally of male and female, or adult and immature. The male or adult is usually on the top or left. Other captions are explained in the text.

The species descriptions follow the American Ornithologists' Union's *Check List of North American Birds, 7th edition* (1983) for the taxonomic order, as well as for the scientific and common names of Florida species. There have been a few changes to names based on recent researches. Each species

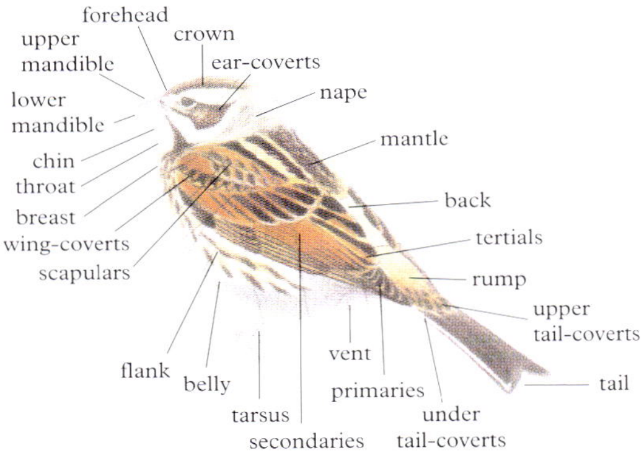

description begins with the common name, scientific name and length of the living bird from bill tip to tip of tail. The next few sentences describe the main identification features. This is followed by a paragraph detailing range, habitat, voice, food, behavior, and other features pertaining to the species. A glossary of terms used in the descriptions is given on page 138, and the diagram on page 5 shows the names applied to the different parts of a bird.

Birding habitats in southern Florida

The subtropical region of southern Florida encompasses ecosystems as varied as anywhere on Earth. It includes pine and deciduous oak forests, mangrove and cypress swamps, scrub, hammocks, coastal prairie, the wet, freshwater marshes and prairies of the Everglades, salt marsh, open salt water, coastal beach and shorelines. There are also large agricultural areas and rapidly expanding urban areas, which annually claim hundreds of acres of natural areas.

The habitats of southern Florida support many species of birds not found elsewhere in North America. These include the Snail Kite, Short-tailed Hawk, Limpkin, Antillean Nighthawk, Smooth-billed Ani and Florida Scrub-jay (Florida's only endemic species). An endemic subspecies, the Cape Sable Seaside Sparrow is restricted to the Everglades. In addition, Florida Sandhill Cranes are smaller than their migratory counterparts from further north and the Red-shouldered Hawk is a smaller and paler bird than its northern form.

The Dry Tortugas support a number of species that are only casual visitors to the mainland, such as Sooty Tern, Brown Noddy and Masked Booby. Uncommon visitors to this area include White-tailed Tropicbird and Black Noddy.

Florida shares some unusual species with other southern states; the Crested Caracara also occurs in south Texas and Arizona, while the white color phase of the Great Blue Heron is found sparingly only along Gulf Coast state coasts.

Many species that breed in boreal regions winter in southern Florida. These include ducks, shorebirds and passerines. Migrants are much in evidence in spring and autumn. In the former season they congregate within coastal South America, before continuing north through the West Indies and onto the Atlantic coast of North America; this flow is reversed in the autumn. The Dry Tortugas are particularly exciting to visit at these seasons, but especially in spring.

Many introductions of exotic species have occurred into southern Florida over the years. As many as 175 species have been deliberately or accidentally released, many as a result of the pet trade. Some simply escaped their confines, while others were liberated during natural disasters such as hurricanes

Key to corner tabs

or floods. Only a handful of these have survived and become established as breeding species.

Pine forests

Pine forests occupy the highest and driest areas of southern Florida. The coastal ridges where they occur are only a few feet or so higher than surrounding areas, but they do not flood. Most such areas have already been exploited for homes and citrus plantations, and little remains except in preserves.

Original pine forests have been logged extensively and have become fragmented in southern Florida; most are in preserves. The original understorey was Saw Palmetto and Wiregrass. Longleaf Pine has been replaced by the faster-maturing Slash Pine in cutover areas. The original understorey has then been replaced by oak and other woody vegetation, which without periodic burning, eventually succeeds into oak scrub. This is the area in which the Brown-headed Nuthatch and the endangered Red-cockaded Woodpecker are likely to be found.

Oak scrub

Oak scrub is widespread in southern Florida, but is the most endangered plant community due to destruction by urbanization and the citrus industry. The remaining areas are fragmented, sparse and stunted. Almost half of the scrub plants are endemic. As an ecosystem it is dependent on fire, and as a result many areas are disappearing due to plant succession. The Florida Scrub-jay is resident here.

Hammocks

Hammocks are islands of tropical hardwoods that form on layers of rock in wetlands or on ridges of limestone in pinelands. In the Everglades, they are higher areas, which resemble teardrops from the air. The rounded edge is upstream, and faces the water flow. Over many years, accumulations of silt form the tail. Accumulation of organic material forms soil. Pineland hammocks are circular. Hammocks are lush compared to the surrounding vegetation.

Cypress swamps

Cypress swamps are the tropical 'jungles' of southern Florida. Bald Cypress trees are draped with Spanish Moss; orchids attach themselves to their branches. The crystal clear waters beneath the trees reveal fishes, alligators, snakes and turtles. Ferns cover fallen trees. The trees are tallest where the water is deepest, and denser along the water's edge. Most of these areas have been drained and logged, and only remnants survive. This was once the home of the Ivory-billed Woodpecker and Carolina Parakeet. Now Swallow-tailed Kite, Barred Owl, Pileated Woodpecker, Wood Duck, Limpkin and Anhinga are common. Wood Stork rookeries are a feature in these areas.

Interior prairies

Interior prairies are typical of the areas around Lake Okeechobee. The soil is sandy and is characterized by potholes and marshes. Thickets of palmetto and scrub oak dot this type of landscape. During the dry summers fires may spread across the prairie lands, which are very open. Birds characteristic of such areas include Sandhill Crane, Crested Caracara and Burrowing Owl.

Wet prairies or Everglades

Wet prairies or Everglades are the freshwater marshes that are flooded either seasonally or permanently. The Everglades is a large freshwater marsh which has been reduced to less than half of its former size due to various ill-conceived land use practices. The Snail Kite and Cape Sable subspecies of the Seaside Sparrow are confined to this area.

Estuaries

The shallow portions of coastal rivers or wetlands that are affected by tides are called estuaries. Fresh and salt water mix in varying proportions here. This area bridges the glades and the mangroves. The surrounding mudflats provide feeding grounds for myriads of water birds.

Mangrove swamps

Mangrove swamps are coastal and are found worldwide in the tropics. Mangroves are the first line of nature's defence, and protect the coasts by absorbing the fury of tropical storms and hurricanes. The mangrove tree is a saltwater plant; three species inhabit southern Florida, with each species occupying a separate niche depending on the depth of the water and the exposure to flooding. The size of the plant increases in proportion to the salt content of its environment. White-crowned Pigeon, Mangrove Cuckoo, Gray Kingbird, Black-whiskered Vireo, and Prairie and Yellow Warblers are resident here.

Coastal shoreline

Coastal shoreline encompasses the beaches and offshore waters while bays are protected by barrier islands or coastal projections. This is a feeding area for a large variety of water and shore birds.

Coastal prairies

Coastal prairies are characterized by Spartina, which grows in large clumps. Cactus, yucca, and other low-growing, salt-tolerant species are abundant on the salt pans. Hurricane debris commonly occurs in these areas. The coastal prairie is the prime area for Seaside Sparrow, and Clapper and Black Rails.

Open sea

The open sea includes the Gulf of Mexico, the Atlantic Ocean and the Caribbean. This is the home of pelagic species, such as boobies, shearwaters, petrels and other seabirds.

The Keys

The Florida Keys are a long series of islands that sweeps south-westward from Miami to the Dry Tortugas. Key West lies 200km (125 miles) south of the eastern end out into the Gulf of Mexico. The Dry Tortugas extend another 110km (68 miles). The Keys were once a series of coral reefs. Originally the upper keys had vegetation that was similar to the Bahamas, while that of the lower keys resembled that of Cuba.

Urban areas

Urban and suburban land is growing in area at an astronomical rate. The landscapes have taken on an exotic atmosphere as most of the landscaping is done with introduced tropical varieties of plants. These plants fruit, flower and seed all year long. This has not only encouraged the native bird population to move into this environment, but has also assisted introduced species' populations to grow and prosper. Exotic parrots, parakeets, mynas, bulbuls and orioles have bred and continue to expand their ranges. Pranty in *A Birder's Guide to Florida* (1996) outlines eleven reasons why southern Florida is an ideal habitat for exotic birds:

1. Subtropical environment.
2. Exotic vegetation.
3. High demand for cage birds.
4. A transient human population that tends to release cage birds rather than move them.
5. A major port of entry for legally and illegally imported birds.
6. A thriving and unregulated aviculture industry.
7. Hurricanes, which destroy bird cages, but not their occupants.
8. Many zoos and roadside tourist traps with extensive bird collections.
9. Extensive suburban feeding area with many bird feeders.
10. A human population which is protective of exotics.
11. Lack of eradication programmes by government agencies.

Agricultural land

Agricultural land is used to produce winter fruits and vegetables, and for citrus growing. These lands support a varied bird life, especially open country species.

Where to find birds in southern Florida

The map on pages 12–13 shows some of the best places to see birds in southern Florida.

Gulf Coast

Punta Gorda

Ponce de Leon City Park

Charlotte Harbor State Preserve

Fort Myers and Sanibel Island
Lakes Regional Park
Ding Darling National Wildlife Refuge
Six Mile Cypress Slough Park

Naples
Rookery Bay National Estuarine Research Reserve

Gulf Coast inland sites
Fakahatchee Strand State Preserve
Corkscrew Swamp Sanctuary
Tamiami Trail
Big Cypress National Preserve
Shark Valley entrance of Everglades National Park

Southern tip and keys
Everglades National Park
Florida Keys Wild Bird Center
Boot Key
Key Deer National Wildlife Refuge
Sugarloaf Key
Key West Botanical Garden
Dry Tortugas National Park

Atlantic Coast
Johnathan Dickinson State Park, Hobe Sound
Hobe Sound National Wildlife Refuge
John D. MacArthur Beach State Park, West Palm Beach
West Palm Beach Solid Waste Authority Facility
Arthur R. Marshall Loxahatchee National Wildlife Refuge, Boynton Beach
Spanish River Park, Boca Raton
Hugh Taylor Birch State Recreation Area, Fort Lauderdale
John U. Lloyd Recreation Area, Dania

Miami area
Greynolds Park, North Miami
Miami Springs
Crandon County Park, Key Biscayne
Matheson County Park
A. D. Barnes Park
Fuchs Park

All of these areas, and many more are described in detail in Bill Pranty's *Birders Guide to Florida.*

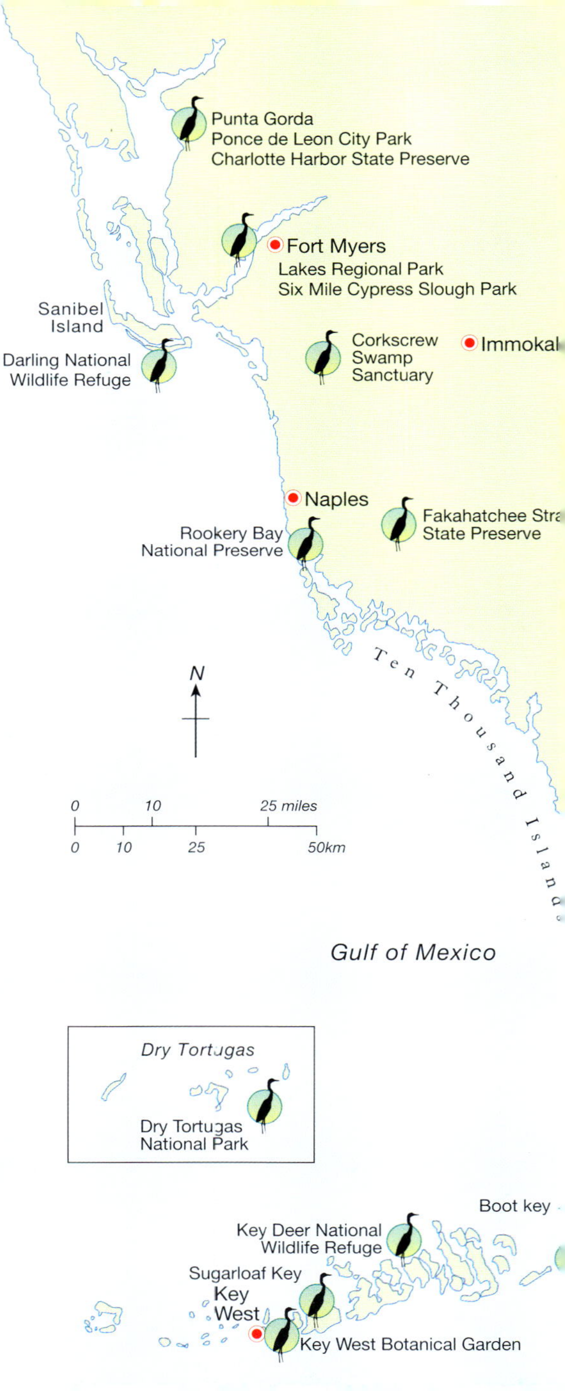
Punta Gorda
Ponce de Leon City Park
Charlotte Harbor State Preserve
Fort Myers
Lakes Regional Park
Six Mile Cypress Slough Park
Sanibel Island
Ding Darling National Wildlife Refuge
Corkscrew Swamp Sanctuary
Immokal
Naples
Rookery Bay National Preserve
Fakahatchee Stra
State Preserve
Ten Thousand Island
N
0 10 25 miles
0 10 25 50km
Gulf of Mexico
Dry Tortugas
Dry Tortugas National Park
Boot key
Key Deer National Wildlife Refuge
Sugarloaf Key
Key West
Key West Botanical Garden

Johnathan Dickinson State Park, Hobe Sound
Hobe Sound National Wildlife Refuge
Lake Okeechobee
Clewiston
West Palm Beach
Loxahatchee Wildlife Refuge
Spanish River Park, Boca Raton
Hugh Taylor Birch State Recreational Area
Fort Lauderdale
Cypress
tional Preserve
John U. Lloyd Beach Recreational Area
Atlantic Ocean
Tamiami Trail
Miami
Shark Valley entrance of Everglades National Park
Biscayne Bay
Everglades National Park
Homestead
Mahogony Hammock
Key Largo
Straits of Florida
Florida Keys
Florida Keys Wild Bird Center
Islamorada
Marathon
indicates best birding area

Common Loon *Gavia immer* 71–89 cm

The Common Loon is an uncommon, but regular winter visitor along both coasts of southern Florida, where it is usually observed in its winter plumage. The bill is pale and has a dark ridge along the upper mandible; its dark crown and nape blend into white underparts. The loon is a large heavy water bird with a large dagger-shaped beak. To take flight it must run along the water surface; when landing it glides along the surface with wings spread. Appears humpbacked in flight. Feeds on fish which it obtains by diving.

Pied-billed Grebe *Podilymbus podiceps* 30–38cm

A grayish-brown duck-like bird, with a conical bill encircled by a black band, can certainly be identified as this species. It is usually observed in pairs or small groups. Common year-round, with the largest numbers reaching Florida in the winter. It inhabits fresh or salt water, but resident birds construct floating nest platforms exclusively in freshwater marshes. An opportunist, it feeds, by diving, on a wide variety of aquatic fare. Dives unobtrusively when disturbed.

Horned Grebe *Podiceps auritus* 30–38cm

In Florida, this species is usually observed in its winter plumage, as shown here. The dark crown and nape, with white cheeks, throat and breast are diagnostic. The bill is slender and the eye red. A fairly common migrant and offshore visitor that also occurs in large bays. Most common in Florida Bay and along the Gulf coast. This species is usually found alone or in very small numbers. Relatively tame and a weak flier. Feeds on small fish and other aquatic animals.

Audubon's Shearwater *Puffinus lherminieri* 27–33cm

Bruce Hallett

This is the only shearwater to occur regularly off the Florida coast. Given a good view, its small size coupled with the dark blackish-brown upperparts and undertail-coverts are distinctive. It also has white underparts and a rounded tail. Rapid wingbeats, between glides, are also characteristic. A marine species, it occurs throughout the warmer oceans of the world, where it usually remains in close proximity to its breeding sites. Feeds on fish, squid and crustaceans.

White-tailed Tropicbird *Phaethon lepturus* 71–81cm

Bruce Hallett

Known as the 'bosun bird', this is a spectacular species. Snow white, with a heavy, yellowish-orange bill, the central tail plumes exceed the overall body length and, in flight, heavy black bars on the upperwings are evident. Usually observed in soaring flight, it is a pelagic species that disperses up to 1000km following the breeding season. Feeds on flying fish and squid by plunge diving, and is among the most acrobatic of birds. Rare in southern Florida, and uncommon in the Dry Tortugas.

Masked Booby *Sula dactylatra* 81–91cm

Bob Starr

Predominately white, with a black tail, primaries and secondaries. The bill is yellow with some black feathering at the base. Gannets are noticeably larger. It is a permanent, but local, resident in the Tortugas, where it perches on buoys and markers. It has attempted, unsuccessfully, to breed in these islands. The Caribbean range of the species is declining due to loss of nesting habitat. It has vanished from former breeding sites. Feeds on larger fish than the other boobies. The Masked Booby of the Pacific is now the Nazca Booby, a new species.

Brown Booby *Sula leucogaster* 71–76cm

W. Vogt

The adult's upperparts are dark chocolate-brown, sharply demarcated from the white abdomen. The bill is pale yellow. Boobies do not spend the night at sea, but return to their roosts. On the breeding grounds, they cradle their single eggs on the feet, while incubating, to transmit body heat, as it has no brood patch. Feeds close to shore on squid and flying fish, which it obtains by plunge diving. The most common booby off the Florida coast, and the one most likely to be seen from land, but rather uncommon in the Tortugas.

Northern Gannet *Sula bassanus* 87–100cm

At a glance, this species could be mistaken for Masked and immature Brown Boobies, but is significantly larger. The wingtips are black, but the plumage is otherwise pure white. Juveniles have dark brown heads and upperparts that become paler with age, but do not attain adult plumage for four years. Plunge dives for schooling fish from great heights. A common, pelagic, winter visitor to the Atlantic coast and Gulf waters.

American White Pelican *Pelecanus erythrorhynchos* 158cm

Unmistakable, this is one of the largest North American birds and has a wingspan of nearly 3m. In flight, look for the large, yellow or orange, flat bill, white plumage and black wingtips. Unlike geese, which also fly in V-formations, each pelican within a line flaps alternately, the result is a synchronized wave. Rather than diving to feed, like the Brown Pelican, it forms flocks in shallow water that corral, then scoop up fish with the aid of their large bill pouches. Mainly a migrant and winter visitor, a few remain in southern Florida throughout the year. Primarily a freshwater species.

Brown Pelican *Pelecanus occidentalis* 107–137cm

The Brown Pelican is a tourists' favorite and can be easily identified by large size, brown color and huge grayish bill. It is commonly observed around fishing boats, docks and piers, searching for scraps and handouts, and sometimes becoming a pest. It can dive from great heights, its entire body disappearing below the surface. Small groups fly with synchronous wingbeats, interspersed with long glides. This is a predominantly coastal species, which is rare in fresh water. It breeds in colonies and avoids open sea, but is the most marine of pelicans. Common along both coasts of Florida.

Double-crested Cormorant *Phalocrocorax aurit*

Dark overall coloration, large size, long neck, hooked bill with yellow throat pouch, and its habit of perching with wings spread are characteristics of the Double-crested Cormorant. A saltwater species, it also occurs in deep fresh water. It swims and dives well, feeding entirely on fish in open water. It can slowly sink out of sight if danger is sensed, but must periodically perch to dry its wings in the sun. Flocks fly in V-formations. It is an abundant resident, with numbers being augmented by many northern breeding birds that winter here. Thousands are sometimes encountered in Florida Bay.

Anhinga *Anhinga anhinga* 85–89cm

Male

Female

Also known as a 'darter', this freshwater species has a long thin neck and a dagger-like bill, utilized to impale the fish it pursues. There is a large whitish patch on the mantle and upperwing, and it frequents shallow, calm waterbodies. Like cormorants, to which it is closely related, it must dry its wings intermittently in the sun after fishing forays. It often swims below the surface with only the head and neck exposed, hence another of its names, 'snake-bird'. Frequently soars with neck extended and tail fanned. It is a resident breeder that builds large bulky nests.

ificent Frigatebird *Fregata magnificens* 94–104cm

The 'Man o' War' has a long, forked tail; it soars very slowly on its rapier-like wings, in and around coastal areas. Males are entirely black, with a red throat in the breeding season. Females are white-breasted, and the immature has a white head and breast, separated by a dark brown band. It is the pirate of the sea, relying upon speed and agility to gather food from the ocean surface, especially flying fish and squid, or to rob other species. This is the only species of frigatebird that occurs in Florida, where it breeds and roosts on some of the small keys.

American Bittern *Botaurus lentiginosus* 58–71cm

Any medium-sized, brown-streaked heron with black neck streaks, frozen in place with its bill pointing skyward is this species. Its contrasting dark brown flight feathers are conspicuous while in flight. It is a solitary, ground-dwelling species that only very rarely perches in trees, and is most frequently seen flying over marshes on rather pointed wings. It is a stealthy, deliberate predator of fish, crayfish, frogs and mice. A common winter visitor to freshwater marshes, it occasionally occupies brackish and saltwater habitats as well. The Everglades comprises its center of abundance.

Least Bittern *Ixobrychus exilis* 28–35cm

Male *Female*

The smallest North American heron, it normally secrets itself in emergent vegetation; when flushed it flies fast and weakly for short distances. Usually solitary, it freezes rather than flies when disturbed. Feeds on small land vertebrates as well as aquatic fare. Remains motionless and is well camouflaged, but is unmistakable in flight. Both sexes are characterized by a dark crown and back that, along with the buff-colored wing patches, are distinctive in flight. Uncommon in fresh and saltwater marshes, where it prefers cattails. A permanent resident in southern Florida.

Great Blue Heron *Ardea herodias* 107–132cm

Normal (left); white phase (above)

The largest heron of the Western Hemisphere and thus unmistakable even at a distance. Blue-gray, with a yellowish bill and black head plumes in breeding plumage. An uncommon all-white morph occurs in southern Florida, which is considerably larger than the otherwise similar Great Egret, from which it is also separable by its yellow legs. It feeds upon a wide variety of prey, such as mice, crabs, shorebirds and frogs, as well as fish. This resident nests alone or in small colonies.

Great Egret *Ardea alba* 89–107cm

Abundant in Florida, this species is common virtually throughout the world. A large, all-white, long-necked species that possesses beautiful head plumes during the breeding season. The bill is long and yellow, and the legs are blackish. A dark line, extending from the gape to beyond the eye, is unique among the white herons. It lives and feeds on a wide variety of aquatic life that it obtains in fresh and saltwater marshes, swamps and shallows of lakes.

Snowy Egret *Egretta thula* 51–71cm

A very dainty beautiful species, its nuptial plumage is brilliant white, making it highly conspicuous. Legs are black, the feet bright yellow to orange and the lores are bright yellow, or even red in the high-breeding season. The bill is black and appears almost pencil thin, before tapering to a point. An active, deliberate feeder, it is occasionally confused with the immature Little Blue Heron, but has greenish legs and a thicker bill. The immature Little Blue Heron lacks head plumes. The species was almost extirpated in North America, but has made an amazing comeback.

Little Blue Heron *Egretta caerulea* 56–71cm

The adult is easily identified by its dark gray coloration, but the white immatures are possible to mistake for a number of other species, including the white morph of an immature Reddish Egret (*see* page 24). It is important to study such medium-sized white herons very carefully, as they could be mistaken for Snowy or even a Cattle Egret. Young Little Blue Herons have no head plumes and the legs are overall greenish. When the water dries up in wetlands, this species adapts to various grassland foods. A common species of the interior of Florida.

Tricolored Heron *Egretta tricolor* 61–71cm

This common resident heron is predominantly dark blue-gray above, with buff-tinged back plumes, and a white belly and undertail-coverts. The immature has a chestnut hindneck and wing-coverts. It stands slightly taller than the Little Blue Heron, but appears smaller and slimmer due to its long slender neck. An acrobatic feeder, its normal fare is aquatic insects, small fish, frogs and tadpoles, as well as other small aquatic life, which it obtains in fresh or salt water It is more common in salt marsh and mangrove swamps along both coasts and the Everglades.

Reddish Egret *Egretta rufescens* 69–81cm

Adult

White morph

Intermediate in size between the larger Great Egret and medium-sized herons. There are two distinct color phases: a dark phase, which is grayish overall, with the head and neck reddish-brown, and a wholly white morph. The best field marks are its larger size and the black-tipped bill, pinkish at the base. It generally appears to have a ruffled appearance; its habit of 'dancing' or 'prancing' through shallow water in search of prey is immediately recognizable. It is an uncommon year-round resident of the coastal mudflats of southern Florida.

Cattle Egret *Bubulcus ibis* 48–64cm

Non-breeding (above)

First recorded in the Guianas in the Western Hemisphere, this squat heron feeds, in open country and pastures, on insects and other food disturbed by cattle, hence its name. It has a yellow bill, with orange-buff on the neck, chest and back in the breeding season; as a non-breeder it is pure white, except the yellow bill and dark legs. In all plumages it assumes a stooped posture. It does not compete for food with other herons, but soon takes over their rookeries, becoming the major occupant. Habitat is open fields and pastures, especially in wet areas. Its population is exploding and has overwhelmed those of native species.

Green Heron *Butorides virescens* 40–48cm

Adult *Immature*

Its small size, variable dark coloration, short neck, and yellow to orange legs are distinctive and effectively camouflage this species. Usually solitary and skulking, it can be encountered in any size body of fresh or salt water. It often fishes from low branches or along the water's edge, and sometimes occurs along wooded streams. When flushed, it flies abruptly, emitting a loud squawking call. This small heron, common in Florida, where it nests, occurs throughout temperate North America and into northern South America.

Black-crowned Night Heron *Nycticorax nycticorax* 58–71cm

A medium-sized heron of squat appearance. The adult has a black crown and mantle, with a white face and underparts, and black-and-white head plumes. The immature is a warm brown, boldly speckled with white. In all plumages, only the feet extend beyond the tail. A successful species of worldwide distribution, it is a common resident in Florida. Principally nocturnal, it is most frequently seen at dawn and dusk, flying from the roost to its feeding areas. Feeds in all types of water, whether fresh, brackish or salt.

Yellow-crowned Night Heron *Nyctanassa violacea* 56–71cm

Another medium-sized heron, this is a rather chunky gray bird. The adult is dark gray with black and yellowish-white head markings and plumes, while the immature is pale gray with finer speckling than Black-crowned Night Heron; it also has a heavier bill. In flight the legs extend well beyond the tail. Solitary and nocturnal, it principally feeds on hard-shelled invertebrates, but takes a great diversity of animal life. It prefers saltwater areas, although it feeds and nests in a variety of habitats. Common throughout Florida.

White Ibis *Eudocimus albus* 56–71cm

Adult (above); immature (right)

The adult is unmistakable, being pure white with a bright red face and legs, and a long sickle-shaped bill. There is some black on the tips of the primaries. Immatures are dark brown above, with white underparts and on the rump, but are easily identified by the diagnostic bill. Resident from the southern USA south along both coasts to northern South America, and common in Florida. Feeds on aquatic life, which it probes from the mud. The closely related Scarlet Ibis (*Eudocimus ruber*) of South America is a vagrant that was subject of a failed introduction attempt to southern Florida in the 1960s.

Glossy Ibis *Plegadis falcinellus* 56–64cm

This dark iridescent violet to green wading bird, with overall dark chestnut coloration usually appears all dark at a distance, with a long, thin decurved bill. It is resident in freshwater marshes, especially those of the Everglades, and feeds on insects, crayfish, and even small snakes. As an expanding New World invasive, its population will never approach the numbers of the Cattle Egret, although it is increasing annually. It is an uncommon breeding bird in Florida.

Roseate Spoonbill *Ajaia ajaja* 66–81cm

Aptly named. The bare greenish head, spatulate bill and carmine wing-coverts that grade into pale pink, are distinctive. The legs are red. White immatures have pale pink wings, with yellow legs and bill. Feeds by swinging its bill from side to side in shallow water, sifting organisms from bottom sediment. Feeds primarily on fish, but invertebrates and plant material are important dietary components. Occurs from the southeast USA throughout most of South America, and common in southern Florida, mainly in estuary habitats. This unique species is a tourist favorite.

Wood Stork *Mycteria americana* 102–115cm

The large size, naked all-dark head, long, slightly decurved bill and white plumage make this species easily identifiable, even at some distance. Commonly observed soaring; in flight the trailing wing edges and tail are black, and the feet extend beyond the tail. It occurs in swamps, mangroves, mudflats and man-made waterbodies, feeding on all types of aquatic animal matter. Its largest rookery is at Corkscrew Swamp. Occurs from the south-east USA to northern Argentina.

Black Vulture *Coragyps atratus* 56–68cm

This carrion feeder is common in Florida. Smaller than the Turkey Vulture, from which it is distinguished in flight by its short wings with white 'windows' at the base of the primaries, and a stubby tail that scarcely projects beyond the wings. The head is black. It is usually seen soaring in tight circles, flapping its wings four to five times before soaring again on flatter wings than those of Turkey Vulture. Often seen along roadsides searching for roadkills. Less common than the Turkey Vulture in areas where they coexist. A resident breeder, except in the keys.

Turkey Vulture *Cathartes aura* 68–80cm

Its large size, bare reddish head, and silver-gray and black underwings are diagnostic. It soars in a broad V (dihedral), and seldom flaps its wings. This is one of nature's sanitarians, feeding on carrion, especially roadkills, and occurring in large numbers at dumps and landfills. It has well-developed vision and olfactory senses. Forms large roosts; at dawn the birds spread their wings to absorb solar heat. Soars in flocks, especially on cloudy days. The bare head enables it to probe the interior of a kill without soiling its feathers. Ranges throughout the Western Hemisphere and is a resident breeder in Florida.

Greater Flamingo *Phoenicopterus ruber* 107–122cm

Unmistakable, even at great distance, these large pink birds filter feed with their heads immersed in shallow water, straining plankton and invertebrates from the silt. Their nests are constructed in shallow saltwater bays, where a single egg is laid on top of a cone of mud. The young are white and look like goslings. Localized breeding populations are known from Central America and the Caribbean, and small numbers reside in Florida Bay. Nesting has not occurred recently. Flamingos are good swimmers; they fold their legs, and paddle with webbed feet.

Fulvous Whistling-duck *Dendrocygna bicolor* 46–51cm

The long neck and tawny-brown plumage of the Fulvous Whistling-duck is distinctive; the sexes are alike. Gregarious, it is commonly seen in open fields, standing in an upright position. In flight the legs extend well beyond the tail. It is more diurnal in its habits than other species of whistling-ducks. Its worldwide distribution is remarkable; it occurs from the southern USA to South America, in Africa south of the Sahara, Madagascar, India and Burma, but exhibits no plumage variation within this discontinuous range. It is resident in southern Florida.

Wood Duck *Aix sponsa* 43–51cm

Male (top); female (bottom)

The gaudiest duck in the Western Hemisphere. The drake has a spectacular facial pattern and green crest, while the bill is orange-red and the breast is chestnut spotted with white. The female has soft brown underparts spotted with buff, and a prominent white eye-ring. It appears long-tailed in flight and is distinctively shaped. It inhabits ponds and riverbanks in densely wooded areas. Nests are located within holes in large trees; the ducklings leap to the ground from great heights. Ranges through east and west coastal North America. It is a resident and winter visitor to southern Florida.

Gadwall *Anas strepera* 48–58cm

The breeding male has a finely scalloped gray body, while the head is brown and the undertail-coverts black. There is a white speculum, which is diagnostic in flight. The female and eclipse male are brown overall, but retain the white speculum. One of the plainest ducks, its wing pattern becomes wholly distinctive in flight. Feeds in fresh and brackish water marshes. Up-ends to take food from shallow bottoms, but also dives frequently. It is an uncommon winter visitor to southern Florida.

Male (top); female (bottom)

David Tipling; Windrush Photos

American Wigeon *Anas americana* 46–56cm

Also known as the Baldpate, the drake has a white crown and belly, with a pale blue bill and a green surround to the eye; the female is warm brown with a grayish head, white belly and blue bill. In flight, the male exhibits a white forewing and black speculum, while the female's upper wing is pale brown, but also has a black speculum. This is a common and widespread North American breeder that moves south from its breeding areas to winter in freshwater marshes, where it is common in southern Florida.

Mallard *Anas platyrhynchos* 51–71cm

Perhaps the most immediately recognized duck. The green head, bright yellow bill, white neck ring and brown breast of the drake are distinctive; the female is largely warm brown, with an orange bill. It is a very adaptable species, being common in city parks and almost all bodies of calmer water. It feeds on a wide diversity of vegetation, including seeds, plants and submerged roots, along with aquatic invertebrates. Although widespread in both North America and the Old World, it is rare in southern Florida.

Mottled Duck *Anas fulvigula* 50–56cm

This female Mallard lookalike has a pale head with an unmarked throat, yellow bill and a white bar to the trailing edge of the speculum; the sexes are alike. It nests in freshwater marshes, but gathers into flocks that may occupy pastures and salt marshes once the nesting cycle is completed. A fairly common resident, it ranges along the entire Gulf coast and most of Florida into north-east Mexico.

Blue-winged Teal *Anas discors* 38–40cm

Attractive dabbling duck. The drake possesses a unique white crescentic mark, which curves in front of the eye, while the duck is warm buff, spotted with dark brown. In flight, both sexes have a paddle-shaped, pale blue inner forewing. Breeds in temperate North America, migrating as far south as northern South America. Feeds on vegetation in almost all types of waterbodies. An abundant migrant and winter visitor to the marshes, ponds and lakes of southern Florida.

Northern Shoveler *Anas clypeata* 43–53cm

Male (top); female (bottom)

The huge spoon-shaped bill is diagnostic in all sexes and plumages. The male has an iridescent green head, white breast and rufous underparts, while the female and eclipse male are warm brown. In flight it displays a pale blue inner forewing, not unlike the Blue-winged Teal, but is easily distinguished by the bill size. The bill has lamellae along its edges that enable it to sieve food from within mud. It up-ends to feed, usually in shallow lakes and marshes, and is a common winter visitor to coastal regions of south.

White-cheeked Pintail *Anas bahamensis* 38–48cm

Also frequently known by its other name of Bahama Pintail, this species is easily identified by the red bill, contrasting white cheeks and its overall reddish-brown coloration. Inhabits mangrove swamps and pools of brackish or salt water, rarely fresh water. It is resident throughout the West Indies where it is a declining, being threatened by overhunting and destruction of suitable habitat. It is the West Indian form that occasionally strays to southern Florida; a smaller form occurs throughout most of South America.

Northern Pintail *Anas acuta* 51–56cm

A distinctive, elegant duck. Both sexes sport a long pointed tail in flight and on the water. The drake has a chocolate-brown head, neck and hindneck, and a white breast extending to and narrowing on the neck. It has gray underparts, and a buff and black ventral region; the female is fawn-brown. Its worldwide distribution includes the Americas, Eurasia and North Africa. A freshwater species that occasionally inhabits coastal saltwater lagoons. Upends to feed on marsh vegetation. A common winter visitor to southern Florida.

Green-winged Teal *Anas carolinensis* 33–39cm

The smallest of the North American dabbling ducks. The head of the drake is reddish-brown with a large green eye surround; there is a white vertical bar in front of the folded wing. Females and eclipse males, in flight, appear tiny and all-dark, with a green speculum. In contrast, in flight, the Blue-winged Teal is paler in color and flashes a light blue forewing. A fast flyer that is acrobatic on the wing. Prefers heavily vegetated shallow-water ponds. It is a migrant and wintering species in southern Florida.

Male (top); female (bottom)

Canvasback *Aythya valisineria* 48–56cm

An uncommon winter visitor to southern Florida that favours brackish water. Adult males have rusty-red heads, while the breast is black and the back and belly white. The sloping head profile and black bill, as well as the whiter back combine to distinguish it from the Redhead (*Aythya americana*); the female is paler and greyer than the latter species, with the same distinctive head profile as the male. Feeds almost exclusively on submergent aquatic vegetation. It has rapid wingbeats and is the fastest flying large duck.

Male (top); female (bottom)

Ring-necked Duck *Aythya collaris* 40–46cm

Male (top); female (bottom)

Although the drake appears black and white at a distance, in favorable light iridescent purple and green can be distinguished, as well as a brownish neck ring. The upperparts and rear are black, and the off-white flanks are separated from the breast by a large, bright white crescent. It has a white bill ring. The brownish female also possesses a bill ring, as well as a pale eye-ring and stripe leading back from the eye. Dives for food and 'runs' across the water before taking flight. Its habitat includes open fresh and brackish water. A common winter visitor to Florida, where the largest numbers occur in the north.

Lesser Scaup *Aythya affinis* 38–46cm

The drake is gray-backed, with a dark purple head and breast. It can be confused initially with the Ring-necked Duck, which has a black back. The brown female has the flanks barred and a white face. Overall similar to the Greater Scaup (*Aythya marila*). In flight, the Lesser Scaup shows an incomplete wingbar, whereas in Greater this extends to the wingtip. Frequents coastal lagoons, lakes and estuaries, as well as large bodies of fresh water. A common winter visitor to southern Florida, where it was formerly abundant.

Hooded Merganser *Lophodytes cucullatus* 41–49cm

One of the most beautiful of all wildfowl, the thin bill and round-crested head create a distinct profile. The head of the male is glossy black, with a white patch that originates behind the eye. The reddish-brown rounded head and thin bill are diagnostic of the female, which has a dull charcoal back. Feeds extensively on small fish, but has been known to take a diverse variety of food, including acorns. It is a winter visitor to freshwater ponds throughout southern Florida, in small numbers.

Male (top); female (bottom)

Red-breasted Merganser *Mergus serrator* 51–64cm

The drake is a ragged-crested, green-headed duck with a long thin serrated bill, a white collar and a reddish streaked breast. The female and eclipse drake are also crested, with a brown head and gray body. Both sexes have red bills and extensive patches of white in the wings. Frequents open bays and lagoons, in close proximity to the ocean, where it feeds on fish, crustaceans and aquatic insects by pursuing them underwater. A common winter visitor to southern Florida.

Ruddy Duck *Oxyura jamaicensis* 35–43cm

Known as a 'stifftail' owing to the propensity to which this, and closely related species cock their dark brown tails upright. This is a small diving duck: the breeding drake is chestnut, with a dark crown and nape, a white face and pale blue bill. Female and winter males are largely pale brown, with a darker brown cap and white on the face. Frequents deep bodies of fresh water and brackish lagoons, with emergent vegetation. Feeds on vegetation sifted from bottom mud, as well as miscellaneous animal matter. It is a fairly common winter visitor to brackish and freshwater areas of southern Florida; numbers vary annually.

Osprey *Pandion haliaetus* 53–61cm

Virtually cosmopolitan, being absent only from the tundra and the Antarctic. Its basic requirements are water and fish. Eagle-sized, the head and neck are largely white, with some black patches, especially on the nape, as well as a well-delineated eye-stripe. The white underparts are occasionally lightly streaked. When soaring, the Osprey kinks the wings at the 'wrist', rendering it unmistakable in flight. Nests are huge and added to every year. The Florida race is entirely white-headed and wholly white below.

Swallow-tailed Kite *Elanoides forficatus* 60cm

One of the most graceful fliers, it is easily distinguished by its striking black and white plumage, and long, forked swallow-like tail. Locates prey while soaring above the canopy, taking lizards, snakes, birds and large insects, which are then eaten on the wing. Gathers into flocks following breeding in order to migrate to South America. A locally common breeding visitor to cypress and pine woodlands of southern Florida, where it prefers areas near water and is present from February until September.

Snail Kite *Rostrhamus sociabilis* 43–48cm

Kevin Karlson

Kevin Karlson

Female (above); male (right)

Distinctively shaped raptor of wetland haunts. Males are slate-gray with red legs; the wings are black, tail-coverts and tail white, with a broad black subterminal band. Females are dark brown and have a white eyebrow. Juveniles are similar to females, but have white-streaked underparts. Feeds upon large aquatic snails, which it gleans from the vegetation, as it maneuvers slowly over marshy areas in active flapping flight. It perches to feed, patiently waiting for the snail to emerge from its shell. Its extensive range extends from southern Florida throughout South America.

Mississippi Kite *Ictinia mississippiensis* 36cm

Gregarious, often hunting in small groups and nesting colonially. Perched birds show long pointed wings, a small rounded head and tiny bill; the eye is red and legs yellow. The adult is largely gray, but the head grades to white and it has a black square-tipped tail. It glides and circles on flat wings that resemble the Purple Martin. Feeds almost exclusively on insects, but occasionally takes small snakes and lizards in the canopy. It is rare in southern Florida, where it is observed, in the southern Everglades, on migration to South America; it breeds in northern Florida.

Bald Eagle *Haliaeetus leucocephalus* 75–108cm

Adult

Juvenile

The national bird of the USA is easily recognizable in adult plumage by its white head and tail, and otherwise dark brown coloration. Immatures are entirely dark brown. The only eagle in southern Florida, it is substantially larger than the Osprey, which it harasses to seize freshly caught fish. Feeds almost exclusively on fish and carrion. Fairly common in our region, where it is a coastal resident and breeder. More northerly breeders also winter here.

Northern Harrier *Circus cyaneus* 40–61cm

A. & E. Morris, Windrush Photos

Kevin Karlson

Also known as the Marsh Hawk, this species is commonly seen soaring over the marshes of southern Florida, where it is a common winter visitor. Males are pale gray, females brown. The conspicuous white rump easily distinguishes it in all plumages. It hunts on the wing while gliding above open expanses. The wings and tail are long and slender. Feeds on small rodents and snakes, which it transports to a perch to consume.

Red-shouldered Hawk *Buteo lineatus* 43–60cm

An inhabitant of moist woodlands, where it is a common resident. The southern Florida form is smaller and paler than northern races. The adult is brownish above, with a rufous shoulder patch and breast; the rest of the underparts are white, barred rufous. There are a number of narrow tail bands and pale primary 'windows'. The brown immature also possesses these distinguishing 'windows'. Feeds on insects, crayfish, fish, amphibians, reptiles and birds.

Broad-winged Hawk *Buteo platypterus* 35–41cm

Kevin T. Karlson

Adult (left); sub-adult (above)

Small and chunky, brown above, white below, with the tail broadly banded black and white. Soars frequently and feeds upon insects, reptiles and small birds. North American birds are highly migratory and winter in Central and South America; those wintering in southern Florida are largely immatures. Kettles of hundreds of birds are sometimes seen utilizing thermals as they cruise slowly southward. A common migrant and winter visitor to southern Florida, where it inhabits pines, glades, coastal areas and the keys.

Short-tailed Hawk *Buteo brachyurus* 43cm

Kevin T. Karlson

Adult dark morph (inset); immature (main picture)

A strikingly proportioned raptor. There are two color phases: the dark morph is the only all-dark *Buteo* in Florida; the pale form has white underparts. Both forms possess several tail bands, with the terminal one darker. It soars with its wingtips upturned. Small birds are usually favoured as prey, being stooped on from a great height. Commonly soars with vultures. Uncommon to rare, in North America the Short-tailed Hawk is confined to Florida, where it is resident and breeds; northern and central populations winter in the south of the state.

Red-tailed Hawk *Buteo jamaicensis* 48–64cm

A highly variable large hawk, best distinguished by its broad rounded wings and reddish tail. It is white below with a black breast-band. The immature has a brownish tail and heavier underparts streaking. Feeds on rodents, lizards, snakes, and large invertebrates. It is frequently encountered in most habitats, sometimes within lowland urban areas. The species is widespread from Alaska to Panama, and is a fairly common resident in southern Florida, where numbers during winter are augmented by more northerly breeding birds.

Northern (Crested) Caracara *Caracara cheriway* 50–63cm

A distinctive bird whether perched or in flight. The large crested head, large bill and red facial skin, and black-and-white barred breast automatically distinguish the adults. Soars on flattened wings like a raven. In flight, large white 'windows' are evident near the wingtips. The immature is browner, but the head and bill remain distinctively patterned. Ranges from the southern USA to northern South America. In southern Florida it is uncommon in the prairies and scrublands around Lake Okeechobee. It feeds on birds, lizards and rodents; sometimes observed with vultures feeding on carrion.

American Kestrel *Falco sparverius* 23–30cm

Male

Female

Also known as the sparrow hawk, this is a small, colorful falcon. The male is principally reddish-brown, with spots of reddish color on the otherwise grey crown; the tail is reddish-brown with a black subterminal band, and there are two black facial bars. Females have brown wings, males blue-gray. Feeds in open areas, on mice, small birds, insects and lizards. It is a common year-round resident of southern Florida that occurs throughout the Western Hemisphere.

Peregrine Falcon *Falco peregrinus* 38–50cm

The 'duck hawk', as its former name implies, feeds on waterfowl and shorebirds by diving upon them at tremendous speed and either capturing them on the wing, or striking them at the base of the skull with their talons. Recognized by its dark head and moustachials, slate-gray upperparts, pointed wings and barred underparts, immatures are brown, with buff, dark-streaked underparts and also possess moustachial stripes. The prey is usually retrieved before it falls to the ground. Uncommon on migration in southern Florida, it sometimes overwinters around estuaries.

Wild Turkey *Meleagris gallopavo* 90–120cm

Brad Martin

Easily distinguished by its very large size, naked head, iridescent green and chestnut plumage and long broad tail. A weak flyer, it usually prefers to run from danger. Males are much larger than hens. Roosts at night high in pine and cypress trees. An arboreal feeder on acorns, insects, other seeds and fallen fruit. This, North America's largest game bird, is common in southern Florida, where it is a resident of pine and cypress areas, but absent from the Everglades and keys.

Northern Bobwhite *Colinus virginianus* 25cm

The only quail in the eastern USA. The male is a chunky, dark brown bird, with a white supercilium and throat; in the female these parts are buff. The flanks are streaked with rows of white spots in both sexes. Occurs in coveys of up to 30 during the non-breeding season. It feeds on the ground, on insects, seeds and buds. Common in southern Florida, especially in open-pine woodland and agricultural areas where it is attracted to brushy habitats. A favored gamebird, it is declining elsewhere in North America through habitat destruction.

Clapper Rail *Rallus longirostris* 36cm

Favors salt and brackish water marshes. A gray bird with long, largely dark legs and bill observed within mangroves or salt marsh is almost certainly this species. Feeds upon a wide variety of aquatic invertebrates and vegetation, which it seizes by probing in the silt. Can swim well. In flight its legs dangle as it drops into nearby vegetation. When disturbed, it wades and runs through the shallows and over exposed roots. Occurs throughout coastal USA and South America; fairly common in southern Florida.

King Rail *Rallus elegans* 38–48cm

The freshwater version of the Clapper Rail, this is a reddish-brown bird with the flanks and abdomen strongly banded black and white. The bill and legs are long. Its habitat is freshwater wetlands with tall dense vegetation. A secretive species, which is most active at dawn and dusk. Feeds on invertebrates and aquatic seeds. Occurs throughout eastern North America into the interior of Mexico, and is a common resident in southern Florida.

Virginia Rail *Rallus limicola* 23cm

Adult

Sub-adult

This miniature version of the King Rail is half the size of its relative and more widespread. It is a reddish-brown bird with a bright red, long bill and legs; the flanks and abdomen are banded black and white. The cheeks are gray. When alarmed it takes cover in dense vegetation and it can swim well. It feeds by clambering on emergent vegetation, searching for seeds and insects. This species is active throughout the day; but more so at dawn and dusk. It is a winter visitor to fresh, salt and brackish waters in southern Florida.

Sora *Porzana carolina* 22cm

Usually the easiest rail to see, inhabiting mangroves, marshes, rice fields and other fresh, brackish or saltwater areas. It is sometimes observed walking unconcernedly in open areas. Easily distinguished by its brownish-gray coloration, black face and bright, stubby yellow-green bill. Feeds on seeds of water plants, and invertebrates. Breeds in North America, wintering to northern South America, and is a common migrant and winter visitor to southern Florida.

Purple Gallinule *Porphyrula martinica* 27–36cm

This colorful and obvious species has bright bluish-purple plumage, a bluish-white frontal shield, and bright yellow legs, with long toes. Climbs to the top of taller swamp vegetation, where it roosts during the evening. Feeds on a large variety of vegetation and aquatic life. Occurs from the south-east USA throughout Central America, and most of South America. A common resident of southern Florida, it lives in freshwater marshes, ricefields, and fringes of ponds, channels and other watercourses.

Common Moorhen *Gallinula chloropus* 30–38cm

The yellow-tipped red bill, red frontal shield and white line on the flanks distinguish it from similar species. For the most part it is a very dark duck-like bird with long legs and toes. Inhabits freshwater wetlands with emergent vegetation, and either still or running water. Swims readily with a bobbing-head motion, despite its unwebbed feet, and is usually rather tame. Feeds upon a wide range of animal material and vegetation. Common worldwide, it is an abundant resident in south Florida.

American Coot *Fulica americana* 34–43cm

This is an unmistakable bird. It is generally gray-black with a white bill and frontal shield. Unlike the Purple Gallinule and Common Moorhen, which are smaller-headed, it is generally seen in open water, where it associates with waterfowl and grebes. Feeds on submerged vegetation by diving and dabbling, but will also graze in open fields. A strong swimmer, it can walk well. Its long toes are lined with fleshy lobes that function for running along the surface or diving. Sometimes observed in large flocks, it is widespread in the Western Hemisphere and an abundant species in southern Florida.

Limpkin *Aramus guarauna* 69cm

A large bird with long legs and a bill that is slightly decurved, its general coloration is brown with large white spotting. Long toes enable it to negotiate walking over aquatic vegetation. Feeds for the most part on large aquatic snails, but also takes crustaceans, frogs, lizards and insects. Secretive, it occurs discontinuously from the south-east USA, where it is common in southern Florida, through southern Mexico and extends to much of South America. Known as the 'crying bird', it is often tame when encountered.

Sandhill Crane *Grus canadensis* 100cm

A North American species that frequents marshes, swamp edges, and open or semi-open areas, including the edges of pine woodland and savannas. This all-gray bird is considerably larger than the Great Blue Heron and has a bare red crown. It flies with the neck fully extended; its call is a loud bugle. Pairs 'dance' and call in courtship during breeding season. Tends to remain in small groups when not breeding. Feeds on aquatic vegetation, frogs and insects. Frequents areas that have been recently burned. The Florida form is resident and non-migratory.

Semipalmated Plover *Charadrius semipalmatus* 16–19cm

This plover is the most widespread of all the small plovers, and is fairly common as a migrant and winter visitor to southern Florida. Some non-breeders remain throughout the summer as well. Distinguished by its stubby bill with orange base, muddy brown upperparts, narrow black breastband, orange legs and white wing stripe; the winter plumage is duller. Its flight is swift and acrobatic. Occurs on sandy beaches, interior mudflats, and plowed fields, where it feeds on a large variety of animal material.

Black-bellied Plover *Pluvialis squatarola* 26–34cm

This, the largest of North American plovers, is also known as the Gray Plover. Its stocky build, short bill, grayish-white overall coloration and, in breeding plumage, black underparts readily serve to identify it. In flight the white rump and black axillaries are additional confirming features. Usually seen along the water's edge. Feeds on invertebrates as well as insects. It occurs worldwide, and is a common winter resident of the coast and sandy beaches of southern Florida. A few non-breeders remain during summer.

Snowy Plover *Charadrius alexandrinus* 14–15cm

Tiny and pale, with a discontinuous breast-band that is reduced to dark spots on the side of the neck; its legs are black. Known as the Kentish Plover in the Old World. Beaches of sand, salt or mud attract this species, and it is almost cosmopolitan along the world's coasts south of 50°N. It is largely non-migratory. Feeds on bloodworms, flies, crabs, beetles, small mussels and other marine creatures. Although widespread, it is rare in southern Florida during migration and winter.

Wilson's Plover *Charadrius wilsonia* 18–20cm

Brad Martin

Its alternate name, Thick-billed Plover, is appropriate, for this species is brown above, white below and has a large broad, black bill. The male sports a large, wide black breast-band; that of the female is brown. Prefers coastal beaches and mudflats. Unlike other small plovers, usually escapes danger by running and tends to remain far from vegetation and areas covered by flotsam. Takes a great diversity of aquatic life as it forages at the tideline, where waves cast up a variety of potential prey. Occurs in warmer areas of the Western Hemisphere. Nests commonly in southern Florida.

Piping Plover *Charadrius melodus* 18cm

Ashy above and white below, the Piping Plover closely matches the color of the sandy beaches where it occurs. It has a black breast-band and crown patch during the breeding season; the legs are orange, the bill black. Occupies areas of beach away from the shoreline, recently dredged areas and offshore islands. Breeds in the USA and Canada, where it is now endangered owing to habitat destruction, and human disturbance of its nesting sites. It is a winter visitor to southern Florida.

Killdeer *Charadrius vociferus* 25cm

The most familiar shorebird within its range. A large plover, brown above, white below, with a double black breast-band and a reddish-brown rump, which flashes in flight. The shrill, loud monotonous call resembles its name. Juvenile birds have only one chest band. It occurs in open habitats, usually on agricultural land, savannas, other grassy and wetland areas, and estuaries. It is sometimes encountered nesting in parking lots and construction sites. A Western Hemisphere species, it ranges from sub-arctic forests south to Peru, and is a common breeding and winter visitor to southern Florida.

American Oystercatcher *Haematopus palliatus* 43–46cm

This bright, flashy black-and-white shorebird inhabits rocky coasts where it primarily takes shellfish, crabs, small starfish and other saltwater fare; its bill is used to expose shellfish by prizing or pounding them open. A crow-sized species, it has a bright red, laterally flattened bill; the eyes are orange, encircled by red lids. The legs are flesh-colored. It can swim well when necessary. Ranges from the eastern coast of the USA through Mexico to Argentina; uncommon on Florida's south Gulf coast.

Black-necked Stilt *Himantopus mexicanus* 34–40cm

Only the Greater Flamingo has longer legs in proportion to its overall size. This unmistakable, delicate, black-and-white shorebird is endowed with bright red legs and a slender bill. In flight the long slender wings are entirely black; the rump and tail are pale gray. It is usually observed in flocks in a wide variety of aquatic habitats, of both fresh and salt water. Feeds upon small aquatic items, which it takes from mud in belly-deep water. Common throughout much of the Western Hemisphere, it is a common summer visitor, but uncommon nesting species in southern Florida.

American Avocet *Recurvirostra americana* 40–50cm

Winter (above); summer (bottom)

A largely white and black shorebird with a sharply upturned bill. In breeding plumage the head and neck are pinkish-buff. Feeds by sweeping the bill from side to side, and is sometimes encountered in deep water, due to its ability to swim well. Breeds on alkaline lakes in North America and is a common winter visitor to southern Florida, although some also remain throughout the summer.

Greater Yellowlegs *Tringa melanoleuca* 29–38cm

A large sandpiper with bright yellow legs; its underparts are speckled in spring, pure white in winter. The dark gray-brown back is speckled or streaked with white, while the wings are dark and the rump white. Similar to the Lesser Yellowlegs, but larger. It also has a longer, thicker bill, which may be slightly upturned. Common on mudflats and in shallow wetlands both fresh and saline. It occurs throughout the Western Hemisphere, usually in small flocks, alone or with other species. Primarily a winter visitor, but a few remain throughout the summer. The *tew* call is diagnostic.

Lesser Yellowlegs *Tringa flavipes* 23–28cm

Considerably smaller than the Greater Yellowlegs. The two species do not overlap in size, although this can be difficult to establish in the field unless both species are observed together within their similar ranges. Its bill is also much shorter than the larger species. It is brownish-gray above streaked white, with white underparts and bright yellow legs. Common throughout the Western Hemisphere, on mudflats and shallows, in fresh or salt water. Winters in southern Florida.

Solitary Sandpiper *Tringa solitaria* 18–23cm

Dark upperparts and lower throat and breast, white outer tail feathers barred with black, a white eye-ring and greenish legs serve to identify this species. It tends to bob its tail as it walks, feeding along the edges of rivers, lakes and other freshwater habitats, never in salt water. Well named, as it never occurs in flocks. It ranges throughout the Western Hemisphere and migrates through southern Florida to its wintering grounds in the West Indies, Central and South America.

Willet *Catoptrophorus semipalmatus* 33–41cm

A large shorebird of pale gray overall coloration, with long bluish-gray legs and bill, which is medium length and thickened. In flight the black and white wing pattern is highly distinctive. Widespread in the Western Hemisphere; it prefers mud and sand flats, with saltmarsh vegetation, in bays and estuaries along the coast. A common to abundant resident, and uncommon breeder in southern Florida.

Transition plumage (above); winter (below)

Spotted Sandpiper *Actitis macularia* 18–20cm

This common bird has bright white underparts with black spots, there is a smudge on the side of the neck and the base of the bill is yellow, becoming orange during the breeding season. Easily recognized, in flight, by its stiff rapid wingbeats and white wingbar. Its teetering and bobbing walk is also distinctive. Forages alone, or in pairs, along margins of both fresh and salt waterbodies. Widespread throughout the Western Hemisphere, it is a winter resident in southern Florida.

(Hudsonian) Whimbrel *Numenius (phaeopus) hudsonicus* 38–48cm

Formerly known as the Hudsonian Curlew, this large shorebird is easily recognized by its long decurved bill and bright head-stripes with brownish-gray upperparts. Usually observed in small flocks, which assume a V-shaped formation in flight. Feeds on beaches, mudflats, sandbars and open fields. Crabs are an important dietary constituent, but it also remarkably feeds on butterflies and moths along the beach. Breeding within the arctic tundra, it is an uncommon migrant along the coast of southern Florida, where a few individuals may occasionally remain in summer or winter.

Marbled Godwit *Limosa fedoa* 40–50cm

Another large shorebird, recognized by its very long, upturned pinkish bill with a dark tip. The upperparts are mottled dark brown; the underparts barred reddish-buff. In flight, the wing-linings are cinnamon. Wades into deeper water to feed than other shorebirds, sometimes chasing between the waves in the manner of a Sanderling. A common migrant and winter visitor to coastal southern Florida, where it is usually found on tidal mudflats, lakeshores and beaches, feeding on fiddler crabs, oysters and insects.

Ruddy Turnstone *Arenaria interpres* 21–25cm

Easily recognized by its distinctive habit of flipping shells and rocks to feed on concealed aquatic items. Common worldwide, it is a chunky, plover-like shorebird. The breeding plumage's harlequin pattern is uncommonly observed in southern Florida, as it only winters in our region; at this season it has dark breast markings and bright orange legs. In flight a distinctive white pattern to the back, tail and upperwings is evident. Lives on mudflats, sandy and rocky coasts.

Winter (top); winter moulting (bottom)

Red Knot *Calidris canutus* 25–28cm

Mike Lane; Aquila

A chunky sandpiper, which appears plover-like. In southern Florida it is an uncommon migrant and winter visitor. During these seasons it is in non-breeding plumage, which is gray above and white below. At all seasons it is distinguished by its white rump, faint wingbar, gray tail and white wing-linings. Some birds remain during the summer when their attractive rusty-red underparts coloration is evident. Commonly observed feeding with both species of dowitcher. Mudflats are preferred at low tide, whereas it gathers on sandy beaches at high water.

Sanderling *Calidris alba* 18–22cm

Unmistakable among regular visitors to beaches. It prefers the breaker areas as it scampers along the shore, skirting the waves and snatching food items, usually insects and crustaceans. It swiftly retreats from an incoming wave in the fashion of a wind-up toy. Its pale gray upperparts and white underparts make it virtually invisible as it races along on short black legs. In flight it displays a bold white wingbar. Widespread worldwide, it is a common migrant and winter visitor to southern Florida; some non-breeders remain throughout the summer.

Semipalmated Sandpiper *Calidris pusilla* 13–16cm

Brad Martin

One of a suite of small sandpipers, collectively known as 'peeps'. These are the smallest of the sandpipers, and the *peep* call note is similar in all species. Commonly known as the 'sand peep' because of its obvious preference for sandy beaches. Distinguished from other 'peeps' by its small size, black legs and generally gray coloration above, with white underparts. Easily confused with the very similar Western Sandpiper due to their similar bill lengths. Usually occurs in large flocks. Widespread in the Western Hemisphere, it is common in southern Florida, except in winter.

Western Sandpiper *Calidris mauri* 14–18cm

Very similar to the Semipalmated Sandpiper, but has a longer, broader based bill, which is usually drooped at the tip. The upperparts are brownish and the underparts white. It usually feeds in deeper water than the Semipalmated or Least Sandpipers. More common along the coast than inland. Feeds on small invertebrates, mainly by probing in soft mud. Widespread in the Western Hemisphere, where it forms mixed flocks with other 'peeps', it also reaches extreme eastern Siberia, wintering south to northern South America. Known as a migrant and winter visitor to southern Florida.

Least Sandpiper *Calidris minutilla* 13–16cm

Tiny and ball-like shape. Brownish above with white underparts, its yellow legs serve to differentiate it from other 'peeps', although beware of individuals with mud-covered legs. It forms flocks with Semipalmated Sandpipers on coastal mudflats and inland freshwater borders, where it feeds upon small invertebrates. Known popularly as the 'mud peep', because of its brown plumage, and due to it usually being encountered on tidal flats and salt marshes. Common on migration and in winter in southern Florida, it is extensively distributed through the Western Hemisphere.

Dunlin *Calidris alpina* 20–23cm

Common shorebird with attractive summer plumage. The long, thick bill with a decurved tip is a useful field mark. In flight it has a bright white wingbar and white rump, separated by a black stripe. Brownish-gray above in winter, when the pale underparts possess smudges of brown on the neck and upper breast. Much more colorful in its breeding plumage of prominently rusty-fringed upperparts and black belly. When feeding tends to submerge its head to feed on submerged insect larvae, mollusks and crustaceans. It appears on mudflats, beaches, and lakeshores. A common migrant and winter visitor to southern Florida.

Short-billed Dowitcher *Limnodromus griseus* 25–30cm

Both species of dowitcher are easily recognized by virtue of their long straight bills, which they use to probe the mud and may immerse completely. Both pump their bills vertically up and down in the manner of a sewing machine. These gray, chunky, medium-sized waders possess yellowish-green legs and, in flight, display a prominent white patch on the back. Most common on coastal mudflats and sandy spits, it prefers bays sheltered from the wind for feeding and is uncommon inland. The series of *tu-tu* notes most easily serves to distinguish it from the Long-billed Dowitcher. A common to abundant migrant and winter visitor to southern Florida.

Common (Wilson's) Snipe *Gallinago (gallinago) delicata* 28cm

Chunky body, large head and very long bill are useful field marks. The head and upperparts are boldly striped with buff, while the brown rump and rusty base to the tail are usually only seen in flight. The breast is buff colored. When alarmed it explodes from a crouch position into the air, then flies erratically before dropping back into the marsh. Feeds by probing for various invertebrates, especially worms. Its habitat is freshwater marshes and glades. A common migrant and winter visitor to southern Florida.

Laughing Gull *Larus atricilla* 38–43cm

Breeding (top); winter (bottom)

Breeding adults have a black hood and outer primaries, white eyelids, a red bill and clean white underparts. In the non-breeding season the head is mottled black on the nape and sides, but the white border on the trailing edge of the wing is distinctive in flight. The dark gray mantle and black wingtips are also diagnostic. The immature is grayish-brown above with a dusky breast. The call is a laughing *ka-ka-ka-ka-ka-kaa-kaa-kaaa*, befitting its name. A common bird of southern Florida, and throughout the tropical and subtropical Western Hemisphere, it is a resident of bays, coastal areas and offshore keys.

Bonaparte's Gull *Larus philadelphia* 33cm

The smallest North American gull. Distinguished, in winter plumage, by the white head with a dark spot behind the eye. The tiny bill is black and has a reddish base. There is a long white wedge on the outer wing. The breeding adult has a black head, with a white crescent behind the eye and red legs. Feeds on small fish taken from the surface of the water. It is an uncommon winter visitor to coasts, ponds and inlets, and also visits flooded fields and freshwater ponds.

Ring-billed Gull *Larus delawarensis* 46–51cm

Adult (top); immature (bottom)

Adults have a yellow bill with a black band, and yellowish legs. Distinguished from the Laughing Gull by its larger size, the absence of a black head in any plumage, and the bill and leg coloration. Similar to, but smaller than the Herring Gull. Occurs from coastal areas and harbors to fields and parking lots. Forages by fluttering, dipping and walking along beaches, mudflats, or fields. An abundant winter visitor to southern Florida. Usually in large flocks that include Laughing and Herring Gulls. Less common at garbage dumps than the latter.

Herring Gull *Larus argentatus* 58–65cm

Adult (top); immature (bottom)

Only the rare Great Black-backed Gull is larger. Adults have gray mantles, while the gray wings have black tips and white mirrors. The bill is yellow with a bright orange spot near the tip, the eye and eye-ring are yellow. Juveniles are dusky-brown with black bills. Feeds on carrion, garbage and fish. Large numbers congregate at docks, public beaches and garbage dumps with other gull species, and bathe inland in fresh water. A common winter visitor along the southern Florida coast, as well as offshore, and the most abundant gull in North America.

Great Black-backed Gull *Larus marinus* 70–78cm

Brad Martin

Much larger than the Herring Gull and the immature plumage is darker. The adult has a black mantle and a very heavy yellow bill with an orange spot near its tip. It is an aggressive, dominant species, which has been expanding its range southward due to the increasing abundance of man-made food supplies. Nonetheless, it is still only an occasional winter visitor to southern Florida, where it usually associates with other gulls in coastal habitats and at dumps.

Gull-billed Tern *Sterna nilotica* 35–38cm

Well named, as it has a broadly thickened bill, as well as comparatively long legs and a rounded head. Most terns have narrow bills, short legs and a crest. Its uniform gray upperparts, extending to the rump and tail, serve to separate it from the Sandwich Tern. The flight is gull-like and, unlike other tern species that dive to feed, it picks its prey from the surface of the water. A cosmopolitan species, being found in most tropical and temperate oceans, in southern Florida it is an uncommon resident of salt marshes, brackish scrub and plowed fields.

Caspian Tern *Sterna caspia* 48–58cm

Brad Martin

Massive and robust, the largest tern in the world, it has a large bright red bill, a black cap with a short crest, and pale gray upperparts. Much bulkier than the Royal Tern; it is a solitary feeder and the least sociable of all the terns, except when it is breeding. Flight strong and swift; hovers before plunge diving for fish. Pirates from other seabirds and has been known to prey on other species' eggs and chicks. A cosmopolitan species, it is a winter visitor to southern Florida where it inhabits both the coastal areas and the interior.

Royal Tern *Sterna maxima* 45–53cm

The large size and bright orange-red bill of this species are diagnostic. The only similar species likely to be encountered is the Caspian Tern, which is slightly larger, has a chunkier, deeper redder bill, with a black band near the tip, and a slightly notched tail. It is the size of a medium-sized gull, and the underside of the primaries is white with a black trailing edge to the tips. The tail is forked, there is a prominent crest at all seasons and a white forehead for much of the year. It occurs on coasts throughout the Americas and is a common year-round resident in southern Florida.

Sandwich Tern *Sterna sandvicensis* 36–41cm

A black-billed, crested tern, having a long slender bill with a bright yellow tip, is certainly this species. The upperparts are pale gray, but the rump and tail are white, whereas the upperparts of the Gull-billed Tern are entirely concolorous. Frequents beaches and bays. Feeds on shrimp, small fish and squid by plunge diving. A permanent resident in southern Florida where it formerly nested. Occurs throughout Atlantic coastal areas of the Western Hemisphere, Europe and Africa.

Forster's Tern *Sterna forsteri* 35–38cm

This species could be confused with the Common Tern, but the upperwings, rump and tail do not provide any contrast, and the upper primaries are silvery-white. The underparts are pearl gray and the tail is slightly longer than in Common Tern. In breeding plumage it has a black cap, orange legs and a black-tipped orange bill. Feeds on small fish by plunge diving. A common winter visitor to the inland waterways and estuaries of southern Florida, where it also visits marinas and around boat docks.

Summer (top); winter (bottom)

Least Tern *Sterna antillarum* 21–24cm

Kevin T. Karlson

The smallest tern in North America. It is a swift, buoyant flier and has a black crown, white forecrown, a yellow bill with a black tip and yellow legs. Feeds along coasts, as well as inland along the banks of rivers. Feeds by plunge diving for small marine items and occasionally takes insects. Highly migratory, it breeds alone or in small, loose colonies, above the high-water line on sandy islands and beaches from North America to Venezuela, but has learned the advantages of nesting on pebbled roofs in urban areas of southern Florida.

Bridled Tern *Sterna anaethetus* 30–38cm

Bruce Hallett

Usually considered a pelagic species, the Bridled Tern breeds on offshore islands close to the coast. It is dark gray-brown above, with white underparts and hindneck. A white stripe that originates at the bill and continues above and behind the eye distinguishes it from the blacker Sooty Tern. The latter displays white only in front of the eye and has a more deeply forked tail. Away from its breeding grounds it is found only at sea, being resident in the tropical and subtropical oceans of the world, and is only seen near the southern Florida mainland in the aftermath of storms, but is regularly observed offshore in the Gulf Stream.

Sooty Tern *Sterna fuscata* 33–43cm

Bruce Hallett

A black and white tropical tern usually seen far from land where it nests on coral atolls. It is a large, heavy-bodied species with a powerful flight; it soars and feeds in flocks over cool currents of the open ocean, where it prefers to take fish from the surface rather than dive. A white streak originates on the forehead, but extends only to a point above the eye, never behind it, and it does display a white hindneck, unlike the Bridled Tern. Deeply forked tail, white wing-linings and white outer tail feathers complete the description. Found in all of the world's tropical oceans, it is perhaps one of the most abundant tropical seabirds. There is a large colony on the Dry Tortugas.

Brown Noddy *Anous stolidus* 38cm

Bruce Hallett

Its compact appearance, overall dark brown plumage, pale white cap and long wedge-shaped tail are distinctive. Sometimes shares breeding areas with Sooty Terns. It does not swim or dive for its prey, but like other pelagic terns takes it from the surface. Diet includes small fish and squid, sometimes taken by moonlight. The Brown Noddy nests on offshore keys and is rarely seen from land. Occurs throughout the tropical and subtropical seas of the world, and nests on the Dry Tortugas.

Black Skimmer *Rynchops niger* 40–50cm

Adult (top & middle); immature (bottom)

Black upperparts and white underparts, but the most striking feature is the black-tipped red bill, which has the upper mandible much shorter than the lower, and the entire bill flattened vertically. While feeding, the lower mandible cuts through the water's surface like a pair of scissors, enabling it to grab small fish and crustaceans. No other Western Hemisphere species feeds in this manner. The flight is swift, agile and graceful. Usually seen in flocks either in flight or loafing on the beach. This common coastal species is observed all-year in southern Florida.

White-crowned Pigeon *Columba leucocephala* 29–40cm

This species is common in southern Florida during the summer, where its numbers increase due to migration of wintering birds from the West Indies; it is uncommon in the winter. It moves freely among the keys, where it prefers wooded areas at lower elevations, especially mangroves and hammocks. It tends to feed and roost in dense cover. Very distinctive in flight, as its conspicuous white crown contrasts with the dark gray body; it is a rapid flyer. Arboreal, feeding mainly on fruits; most active at dawn and dusk.

Eurasian Collared-dove *Streptopelia decaocto* 28–32cm

A medium-sized gray dove, with dark primaries and a black collar on the hindneck is certainly this species. Feeds on the ground and is frequently encountered in urban areas. Introduced into the Bahamas in the early 1970s, since this time it has spread to Cuba and islands of the Lesser Antilles, as well as to Florida. This native of Europe and Asia has subsequently invaded and colonized many areas of the southern USA. It is now very common within its American range.

White-winged Dove *Zenaida asiatica* 28cm

Distinguished by its pale brown plumage, large white wing patch on black wings and long tail with white corners. Its distinctive *who-cooks-for-you* call has many variations and is frequently heard in the early morning. Feeds on the ground, where it associates with Mourning Doves, on seeds and grasses. The White-winged Dove is common in urban and cultivated areas, but less common in the natural habitat of southern Florida. Natural vagrants occurred for many years, but in the late 1950s the species was introduced from a captive population and has since become a common resident.

Zenaida Dove *Zenaida aurita* 25–30cm

This chunky, richly colored ground-dwelling dove has gray-white tips to the outer tail feathers, vivid purplish neck-sides, and a conspicuous white trailing edge to the secondaries and rounded tail visible in flight. A species of open country, especially in lowlands, it also occurs, more sparingly, in woodlands. Feeds on seeds on the ground but will also take fruit from trees. A common permanent resident in the West Indies, as well as the Yucatán coast of Mexico, in southern Florida it is now a casual visitor that nested in the 1800s.

Mourning Dove *Zenaida macroura* 28–33cm

An abundant North American species, this brown dove with a purplish sheen to the upperparts is easily distinguished by its long wedge-shaped tail with white edging, and lack of white in the wings. In dry areas, flocks will fly long distances to drink at dawn and dusk. Feeds on seeds and other plant material foraged on the ground. Usually found in flocks in lowland, cultivated areas and open country, in southern Florida residents are joined by northern migrants in winter. Its range is increasing due to deforestation.

Common Ground-dove *Columbina passerina* 15–18cm

A small dove, which is common in southern Florida and locally abundant in the keys. The male has a bluish-gray crown and hindneck, while the female is smaller and paler. Easily identified in flight by its small size, reddish wing patches and dark stubby tail. It is a seed-eating ground-dweller of open areas. Flies only a short distance when alarmed. Its world range includes the southern USA through Central America and northern South America. Sometimes takes refuge in trees. Can be very tame around civilization.

Male (top); female (bottom)

Key West Quail-dove *Geotrygon chrysia* 27–31cm

This winter visitor comes to southern Florida to seek its favored berries and fruits. The Key West Quail-dove has reddish-brown upperparts and whitish underparts, heightened by the green and purple iridescence to the head, nape, mantle and back. There is a bright white stripe below the eye, which extends from the base of the bill to the nape. It prefers the understory of dense forest, or dense thickets in arid areas, and is often audible walking through the forest. Takes to the wing silently. A casual winter visitor to southern Florida.

Budgerigar *Melopsittacus undulatus* 18cm

The 'parakeet' or 'budgie', as it is known in the pet trade, is an Australian species that was briefly abundant in southern Florida as a result of its introduction into urban areas, especially along the Gulf coast. Most of these small long-tailed parrots are green, marked with yellow and black, but some exhibit domestically bred color characteristics. A white wingbar is evident in flight. Flocks with European Starlings and Red-winged Blackbirds. It poses little threat to native species, except that it is a hole nester, which could present future problems. Common at feeders.

Rose-ringed Parakeet *Psittacula krameri* 40cm

David Tipling; Windrush Photos

This medium-sized, bright green, long-tailed parrot is an introduced Old World species that breeds in the Miami area. It is much larger than the previous species, the bill is reddish and the male has a black chin and pinkish-red collar. In flight the yellow wing-linings are conspicuous. Colonies are localized, and large communal roosts are the norm. It nests in holes. Feeds around gardens and in orchards. A highly adaptable species that has survived well around habitation and agriculture. Constantly screeches and squabbles.

Yellow-headed Parrot *Amazona oratrix* 36cm

(Far right) Hybridized form

Large, with a bright yellow head, red speculum and yellow, or red and yellow, on the leading edge of the wing. The tail is rounded. Feeds on a variety of fruits including citrus, seeds and grain. Nests in cavities. Originally introduced, from Mexico and Central America, because of its acknowledged ability as one of the best 'talkers', it is increasingly uncommon within its native range, due to such exploitation by the pet industry. Several geographical forms of this species, which are sometimes considered different species, have been introduced to southern Florida, where they have freely interbred giving rise to some unusually plumaged hybrids.

Mangrove Cuckoo *Coccyzus minor* 28–34cm

This rather sedentary and unobtrusive species is usually best located by its low guttural *gawk, gawk, gawk* calls. Slender, with a white-tipped tail and a long, decurved yellow-based bill, the black ear patch and buffy abdomen are diagnostic. Feeds on insects especially caterpillars. Its habitat includes mangroves and hammocks. Resident throughout southern Florida, small numbers of migrants also reach the state. Its range extends into the West Indies, Central and northern South America.

Smooth-billed Ani *Crotophaga ani* 30–33cm

The glossy black plumage, heavy, deep-based bill and long, flat tail are wholly distinctive. Occurs in small noisy flocks, which tend to fly on weak, flapping wingbeats, in 'follow-my-leader' fashion. Groups roost by squeezing together on a perch in order to remain in body contact. Nests are built, and used, communally by a number of females; the young are also cared for and fed by the group. Diet includes insects, lizards and frogs. Commonly observed in small flocks feeding around cattle. Uncommon to rare in southern Florida, it was formerly uncommon near Lake Okeechobee.

Barn Owl *Tyto alba* 30–43cm

Its pure white underparts, silent flight and piercing screams have given rise to haunted house legends. The white, heart-shaped face and pale brown upperparts heighten its ghostly nature. Nocturnal, it feeds principally on rodents, large insects and even bats. Perches on fence posts, especially around rice fields, dry scrub and open woodland. Usually observed soaring at dusk. In urban areas they may hunt over shopping malls and sports arenas, and has the capacity to live in close proximity to man. This most cosmopolitan of owls is a fairly common resident in southern Florida.

Eastern Screech-owl *Otus asio* 18–25cm

Strictly nocturnal. This small owl has erectile ear tufts and there are two color morphs that possess either rufous or gray upperparts; the white underparts are streaked with the same color as the mantle and wings. Inhabits woodlands of all types, where it roosts and nests in holes. Feeds on small rodents for the most part, but also takes small birds, insects and lizards. Its whinny or wailing call is very distinctive, and is usually heard soon after nightfall. The screech-owl is a common species in southern Florida.

Great Horned Owl *Bubo virginianus* 45–63cm

This, the largest Florida owl, has prominent ear tufts that are spaced widely apart, a white throat and heavily barred underparts. It is nocturnal, but may hunt on cloudy days; its diet incorporates a wide variety of prey, including domestic cats, skunks and herons. The flight is silent, swift, graceful and powerful. It nests during winter, usually occupying an old hawk or eagle nest for this purpose. It is an uncommon resident in southern Florida, where it frequents pines and hammocks even near human habitation.

Burrowing Owl *Athene cunicularia* 23cm

Mottled brown and white above, with barred underparts, long legs and an upright stance, this species is one of the most charismatic owls. As its name suggests, it lives and nests in underground burrows. Characteristically bobs up and down when it is approached. Occurs in well-spaced colonies. Diet includes insects, spiders, frogs and small birds. Hunts usually at dusk and by night, but also during the day. It is a terrestrial species that occasionally perches on fences. Its habitat, throughout its range, is open treeless grasslands, open scrub and sandy flats. Uncommon in the region.

Barred Owl *Strix varia* 30–50cm

This large-headed owl has no ear tufts, the upperparts are gray-brown, the underparts whitish, streaked brown, the neck and breast are barred, and the eyes brown. Although of large appearance, it feeds on small prey such as frogs, crayfish, small birds and mammals. The repeated deep *who-who-who cooks for you all* vocalization is distinctive, and is likely to be heard during the day as well. Perhaps the most common owl of southern Florida, Barred Owls rely on cypress swamps, hammocks, pines and dense forests for roosting and nesting. This species is nocturnal, and very docile. The roost is well hidden.

Common Nighthawk *Chordeiles minor* 24cm

Brad Martin

In its remarkably jaunty, buoyant flight, the long pointed wings marked with white patches close to the bend are characteristic. Underparts are barred, the tail notched. Nighthawks roost on branches or in brush during the day, when their intricately marked plumage can be fully appreciated. During the evening and around dawn they feed on flying insects such as flies and mosquitos. Their distinctive flight call is a nasal *peent* similar to that of the American Woodcock (*Scolopax minor*). A common migrant and summer visitor to southern Florida, where it is observed soaring over pines, prairies, beaches, agricultural land and even shopping malls.

Antillean Nighthawk *Chordeiles gundlachii* 20–25cm

Similar to the Common Nighthawk, the two species are only separated with some difficulty in the field. This species' flight is more erratic and it has tan wing-linings; its call is *pity pat pit* given on the wing. In contrast, the Common Nighthawk utters a nasal *peent* and the wing linings are black. Roosts on the ground, or lengthwise on a horizontal branch. Principally active at dusk and dawn. Effortlessly takes moths and beetles on the wing. Sometimes occurs in large flocks. In southern Florida it is a summer resident but confined to the keys.

Chuck-will's-widow *Caprimulgus carolinensis* 30cm

Judy A. Elseroad

The largest of the North American goatsuckers, it frequents swamps, hammocks, and pines. Chuck-will's-widow is reddish-brown, mottled and spotted with black and the chin is brown. The male has some white edging to the tail feathers and a white throat patch that can be difficult to observe. The wings are short and rounded. Feeds on large insects. Perches and nests on the ground. Its distinctive call is a four-syllable *chuck-wills-wi-dow*, with the emphasis on the first and third notes. Common in southern Florida, where it is a summer visitor.

Whip-poor-will *Caprimulgus vociferus* 24cm

Kevin T. Karlson

Mottled gray-brown plumage, with black spots and streaks; the chin is black. A medium-sized species, the male has white outer tail feathers, but the female has no white in the tail. Nocturnal, it feeds on insects, which are pursued on the wing. Roosts on the ground or on a horizonal limb during the day. Much smaller than the Chuck-will's-widow. The three-syllable *whip-poor-will* call is repeated in long series', with accent on the first and third notes. A fairly common migrant and winter visitor to hammocks and pines, but easily overlooked in our region because it rarely vocalizes at these seasons.

Chimney Swift *Chaetura pelagica* 14cm

The 'flying cigar', as it is affectionately known, is an uncommon migrant in southern Florida, although a few may summer in the far north of our region. This swift is uniformly brown above, the throat and upper breast are pale. Its tail has a number of stiff projecting spines at the tip. Its distinctive silhouette is usually seen far overhead as it flies swiftly, alternating shallow wingbeats and periods of gliding. An aerial feeder on insects, it is the only swift in the area covered by this book.

Belted Kingfisher *Ceryle alcyon* 28–36cm

Wayne Lankinen; Aquila

This striking species has a blue-gray shaggy crest and upperparts; the male has a blue breast-band on white underparts and the female a double band, the lower one being orange. Feeds on small fish and usually observed on a conspicuous perch. Hovers before diving. Its wingbeats are deep and rapid, but irregular. The distinctive rattling call often draws attention. Frequents streams, shores of lakes and ponds, and coasts. Solitary except during breeding season. The common kingfisher of North America, it is abundant in southern Florida.

Bahama Woodstar *Calliphlox evelynae* 9–9.5cm

Male

Female

A small, aggressive hummingbird that is endemic to the Bahamas, although it occasionally wanders to southern Florida. The male is green on the back and wings, as well as the head, the underparts are white and the throat brilliant magenta. The tail has a shallow fork. Females lack the magenta throat and have rounded tails. Feeds in gardens, wood edges and coppices on nectar and insects, and is also known to hawk for midges. Very tame, it perches on twigs and wires.

Ruby-throated Hummingbird *Archilochus colubris* 8–9cm

Male

Female

This tiny bird is iridescent green above and mainly white below. Adult males have a bright red gorget that in poor light appears black, the sides and flanks have a pale olive cast and the tail is deeply forked. The female's tail is more rounded, with white at the corners. It frequents areas with plentiful flowers, such as parks and gardens. It crosses the Gulf of Mexico to spend the winter in Central America. It is the common hummingbird of the eastern USA, in southern Florida it is usually observed during migration, although some remain to winter. Nesting is occasionally recorded.

Red-bellied Woodpecker *Melanerpes carolinus* 23–26cm

Male

Female

The most common woodpecker in southern Florida, it inhabits pines, mangroves, woodlands and even urban areas. The gray-white base color to this species is enhanced by the barred upper-parts; the rump is white and the face plain. Males have a red crown and nape. The female's head is plainer, but it has a red nape and forehead. The red belly patch, from which it takes its name, is usually obscure. It feeds on insects, nuts, fruits and berries, and sometimes stores food. Common at feeders where it is a very vocal and aggressive species.

Yellow-bellied Sapsucker *Sphyrapicus varius* 19–23cm

Male (left); female (middle); juvenile (right)

Easily identified by its large white wing patches, evident in all plumages. The male has red on the crown and throat; the female on the crown only. The head is white with a broad black eye-stripe, the breast black, and the underparts are pale yellow, speckled with black on the flanks. It drills rings of holes in live pine trees. The sap that accumulates in these small pits attracts insects, and the sapsucker returns to harvest them along with the sap. Inhabits forests, gardens and pines. A common bird in southern Florida.

Downy Woodpecker *Picoides pubescens* 16cm

Male

Female

Its small size, black-and-white plumage, and short bill are diagnostic field marks. The male has a red nape patch, absent in the female. It drills into hardwoods less often and less loudly than the similar Hairy Woodpecker, and its calls are also softer than the larger species. It forages on tree trunks, hanging upside-down to glean insects from the foliage. The smallest woodpecker in southern Florida, where it frequents pines and hammocks; it is also common in urban areas where it prefers gardens and orchards, and frequents feeders.

Hairy Woodpecker *Picoides villosus* 20–23cm

Male

Female

Sometimes confused with the much smaller Downy Woodpecker, this species is principally black above and white below. The male has a red spot on the back of the head, which is absent in the female. The moustachial stripe, as well as the eye-stripe, is black. An inhabitant of diverse woodlands, where it drums loudly on trees. Feeds on the lower trunks of large trees, but is sometimes also observed on the ground. When foraging in pines, males feed on cones, females on the trunks. A significant percentage of its diet consists of harmful insects. Uncommon in southern Florida.

Red-cockaded Woodpecker *Picoides borealis* 21cm

Black-and-white, with a barred back, it also has white cheeks, a black cap and black streaks on the flanks. The Red-cockaded Woodpecker nests in live pines with heartwood disease; holes are bored through the outer hardwood, then downward into the tree. Small holes are then drilled in the area around the nest entrance to ensure a steady supply of resin that repels insects and acts as a signpost. Small groups defend mutual territories. Feeds almost entirely on harmful insects. A resident of the mature pine woodlands of southern Florida, this woodpecker has become endangered due to habitat destruction.

Judy A. Elseroad

Northern Flicker *Colaptes auratus* 30–32cm

In flight the flicker displays bright yellow underwings and a white rump spotted with black. At rest, it is a large woodpecker with a black bar across the breast, pale buff underparts spotted with black and a red patch on the hindneck. The male has a black moustachial stripe. Frequents open forest and edges, but any area with trees near open ground is utilized. Feeds commonly on the ground where it hops about probing for ants, which are its primary food source. Also feeds on fruits, berries and nuts. It is widely distributed throughout North America, including southern Florida.

Pileated Woodpecker *Dryocopus pileatus* 40–49cm

With the probable extirpation of the Ivory-billed Woodpecker (*Campephilus principalis*), the largest woodpecker in North America and southern Florida. Easily recognized by its general black coloration, large size and long red crest. The male has a red moustachial stripe; the female's is black. Its flight is powerful and undulating. Primarily feeds on carpenter ants, but also takes acorns, seeds, wood-boring insects and fruits. Fairly common resident of mature woodlands of many types: pine, cypress and hammocks especially those with an abundance of dead trees. Also encountered in urban areas with large trees, which are required for nest sites.

Eastern Wood-pewee *Contopus virens* 15–16cm

Dark gray-olive above, the throat is whitish, breast and flanks dark gray, and the rest of the underparts yellowish-white. The upper mandible is black, the lower dull orange with a black tip. The most obvious feature is the prominent whitish wingbars. Forages for insects from a perch in the canopy of woodlands. The *pee-a-wee* call is distinctive. An uncommon migrant and rare winter resident in southern Florida.

Acadian Flycatcher *Empidonax virescens* 14–17 cm and
Least Flycatcher *Empidonax minimus* 13–15cm

Acadian Flycatcher

Least Flycatcher

The Acadian Flycatcher is a regular migrant through southern Florida, where it is common in the west, but rare in the east. The Least Flycatcher is an uncommon migrant and rare winter visitor. Both have distinct eye-rings and buffy wing bars. The whitish underparts shade to a brighter yellow on the flanks and abdomen of the Acadian. Since both species appear nearly identical in the field, the voice is the main diagnostic feature. The Acadian has a *peet-seet* song, wheras that of the Least is *chebek*...

Eastern Phoebe *Sayornis phoebe* 16–18cm

Brown above, darkest on the head, the underparts are white and there is a yellowish wash to the flanks and breast in fall plumage. There are no wingbars. Usually observed sallying from a perch to take flying insects. It bobs and spreads its tail continuously. Commonly nests under bridges, and even beneath eaves and rafters. Its *fee-be* call is immediately recognizable. A common migrant and winter resident in areas near fresh water such as swamps, sloughs, canals and ponds, in southern Florida.

Great Crested Flycatcher *Myiarchus crinitus* 20–23cm

Olive-brown upperparts, with grayer throat and breast; the belly is bright yellow and white wingbars are evident. The outer margins of the primaries and the inner webs of the tail feathers are cinnamon. The lower mandible is horn colored at the base. Bushy crest. Frequents pines, hammocks, cypress swamps, mangroves and other types of woodland. Sallies from its perch, usually a dead branch, and captures insects in the air with an audible snapping of its bill. Usually encountered with raised crest and occasionally bobs its head while searching for insects. Distinctively calls *hank-hank*. Common resident in southern Florida.

La Sagra's Flycatcher *Myiarchus sagrae* 19–22cm

A medium-sized *Myiarchus* endemic to the Bahamas, Cuba and the Caymans. Its manner of perching is unique, as it leans forward at a 45-degree angle. The flat-headed appearance, black bill and faint wing-bars are diagnostic. Upper-parts are grayish-olive; the rump is cinnamon during breeding, but gray at other times. The crown is dark, the neck-sides and breast gray, and the underparts grayish-white. It elevates its crest when disturbed. Frequents the understory of pine and mixed wood-lands, at all elevations. Sometimes found in man-groves. Common, except in the southern Bahamas. It wanders to southern Florida, where it usually occurs near the coast.

Western Kingbird *Tyrannus verticalis* 20–24cm

Joe Fuhrman; Aquila

Usually encountered in open fields. The head, breast and nape are gray, the belly is yellow, the back dark gray and the black tail has distinctive, white outer feathers. It perches on utility poles, fences, wires and treetops, usually in full view. Feeds almost exclusively on insects taken in the air. Intolerant of other species that invade its nesting or feeding territory, and commonly mobs much larger birds such as hawks and crows. An uncommon migrant and win-ter visitor to southern Florida.

Eastern Kingbird *Tyrannus tyrannus* 20cm

The entire crown, sides of the face and upperparts of this species are black. A bright white terminal band enhances the black tail. Underparts gleaming white. Common in open areas especially at the edges; flies with shallow, trembling wingbeats. Nests in shrubbery near water. Very territorial and fearless, it commonly attacks and drives away much larger species. It sometimes carries the pursuit for long distances. Feeds almost exclusively on insects, most of which are injurious to crops. A common migrant and summer resident in southern Florida.

Gray Kingbird *Tyrannus dominicensis* 22–25cm

Gray above, white below, with a black mask, which extends from its large bill below the eye to the ear-coverts. The tail is notched. Feeds on large insects, which it catches on the wing, before returning to its perch and pounding the prey on a hard surface to soften it prior to eating. Sometimes feeds at night under streetlights. Commonly seen on wires and the tops of scattered trees. Like other kingbirds it is very aggressive. Occasionally forms large communal roosts. Abundant in the West Indies, it also breeds in the south-east USA and northern South America.

Loggerhead Kingbird *Tyrannus caudifasciatus* 24–26cm

Large flycatcher with dark upperparts and a black crown, which extends below the eye. The underparts are pure white. Square-ended black tail, tipped off-white, except for those in Puerto Rico and Hispaniola. The very large black bill is probably its most noticeable feature. Frequents more forested areas, including dense scrub, than the Gray Kingbird. It sallies for insects and small lizards from low exposed perches, or telephone lines. Endemic to the northern Bahamas, the Greater Antilles and the Caymans, but only accidental in southern Florida.

Scissor-tailed Flycatcher *Tyrannus forficatus* 30–38cm

Mike Wilkes; Aquila

This spectacular flycatcher is basically pearl gray, while the wings, uppertail-coverts and central tail feathers are black. Pale pink on the belly, undertail-coverts, flanks and underwings. The very long outer tail feathers are highly distinctive, even given only a brief view. Usually seen on wires and fences in open country. Feeds on large insects by dropping onto them from a high perch. The scissor-like motion of the tail is unique. A common migrant breeder in the central west USA, it is only an uncommon migrant and winter visitor to southern Florida.

Loggerhead Shrike *Lanius ludovicianus* 23cm

The 'butcher-bird' is a common southern Florida species that is easily recognized by its black mask, which extends over its short hooked bill. The upperparts including the cap are gray, while the underparts are uniform gray-white, except in the juvenile, which are barred. Frequents open habitats, where it perches on low branches, fences and utility wires, scanning for its prey of insects and mice. After a successful capture, the victim is sometimes impaled on a thorn or barbed wire, where it is stored for a future meal.

White-eyed Vireo *Vireo griseus* 13cm

The principal field mark is the yellow spectacle, which encloses the white iris and extends to the forehead. The upperparts are gray-green, the underparts are white, and there are two prominent white wingbars. Inhabits wet thickets in hammocks, mangroves and deciduous woodlands, usually near the ground. It is a loud, constant singer. Feeds mainly on insects, but is capable of switching to a diet of fruits and berries. The local Key West race has a larger bill and is consequently sometimes mistaken for the rarer Thick-billed Vireo. A common resident of various habitats in southern Florida.

Thick-billed Vireo *Vireo crassirostris* 14cm

A small bird that is generally brownish-green above, with a gray-green crown and hindneck. Two white wingbars and the bright yellow spectacles that surround the eyes are probably the most obvious field features. The lores are black and the iris dark. Underparts coloration varies from yellow to gray. Curious and tame, it inhabits dense shrubbery and undergrowth. Usually forages in pairs, feeding on insects and berries in low vegetation. This vireo is a resident of the Bahamas, and is only a casual visitor to southern Florida.

Yellow-throated Vireo *Vireo flavifrons* 13cm

This is the only vireo that possesses a bright yellow throat, breast and spectacles. There are also two bright white wingbars. The rest of the underparts are white, while the upperparts are olive darkening to gray on the back and rump. Frequents hammocks, wooded swamps and semi-open forest where it forages in the upper canopy. Feeds on butterflies and moths as well as other insects. Its common two-syllable song is *three-eight*. An uncommon migrant and winter resident to southern Florida. Can easily be mistaken for Pine Warbler.

Blue-headed Vireo *Vireo solitarius* 13–15cm

Kevin T. Karlson

Its bluish-gray head, contrasting with the greenish mantle, white underparts and yellowish-green flanks make for a reasonably straightforward identification. It has bright white spectacles and two white wingbars. Its movements are slow and deliberate as it feeds, in the canopy or understory, but usually near the apex of the plant, on insects and berries. It appears tame and often permits close approach. This species is a recent 'split', whereby Solitary Vireo is now considered three different species (Plumbeous *V. plumbeus*, Cassin's *V. cassinii* and Blue-headed Vireos), with the latter breeding east of the Rockies and migrating through southern Florida, where it is also an uncommon winter visitor.

Red-eyed Vireo *Vireo olivaceus* 15cm

The dark gray crown, with a black border, whitish supercilium, black eye-stripe and red eye are diagnostic. The upperparts are olive-green and the underparts white, but there are no wingbars, unlike most vireos. Common throughout hammocks, swamps and mixed woodlands along the coast. It is a canopy dweller and rather sluggish in its movements. When singing it is easily recognized by its loud, repetitive song. Feeds almost exclusively on harmful insects. A common migrant in southern Florida, where a few pairs remain to nest.

Black-whiskered Vireo *Vireo altiloquus* 15–17cm

Similar to the Red-eyed Vireo, it has olive-green upperparts, white underparts washed with yellowish-green, a white supercilium, a dark line through the eye and a black malar stripe. There are no wingbars. Its loud distinctive one to four-note song, given throughout the day, readily draws attention to this otherwise rather shy species. Feeds on insects and fruits, which it sometimes takes on the wing. Frequents the canopy of coastal mangroves and hammocks. Nests are placed over water. It is a common summer resident in southern Florida.

Blue Jay *Cyanocitta cristata* 28–31cm

The blue crest and upperparts, black necklace, long blue tail and white underparts are wholly diagnostic. Usually found in small flocks, this jay has a catholic diet of fruits, seeds, acorns and insects. It robs nests, taking both the eggs and nestlings of other species. Its loud *eef-eef* call echoes through forests. One of the most familiar North American birds, the Blue Jay is a common resident of pines, hammocks, citrus orchards and most other wooded areas of southern Florida. It is also widespread throughout urban areas.

Florida Scrub-jay *Aphelocoma coerulescens* 28–30cm

Brad Martin

Adult (left); juvenile (above)

The scrub jay has no crest, and its upperparts are blue and the back gray-brown. The throat is streaked white, grading to gray on the rest of the underparts, and there is a faint bluish necklace across the upper breast. Unlike the two western scrub-jay species, the Florida form has a white forehead and eyebrow. Frequents scrub oak, palmetto and myrtle, preferably in areas subject to burning. It breeds cooperatively and feeds in small groups at ground level, taking acorns, insects and berries. This jay was once common in southern Florida, but its numbers were seriously reduced in the 20th century by habitat destruction.

American Crow *Corvus brachyrhynchos* 43–53cm

This very large all-black bird is a common resident of southern Florida, where it inhabits a wide variety of habitats, including woodlands and wetlands of all types. The long, heavy bill is distinctive. An omnivorous species, and along with a diet of acorns and berries, this crow is a major predator of heron rookeries, taking both eggs and young birds. Usually occurs in small flocks; its *caw-caw* call is a familiar sound throughout its range.

Fish Crow *Corvus ossifragus* 40–50cm

A. Morris; Windrush Photos

Slightly smaller than its cousin, the previous species, both species share an identical black coloration. The two are safely distinguished in the field only on voice, a nasal *cah* in this species, although be aware that young American Crows have a similar call. Unlike its close relative it prefers to pirate the nests and eggs of songbirds, and commonly takes squabs from ledges in larger cities. It also feeds on carrion and garbage, and scavenges among marine debris along the coast. Frequents dumps, islands and rivers. Common in southern Florida, its populations are currently on the increase.

Purple Martin *Progne subis* 18–21cm

One of the earliest spring migrants. Male easily recognized by virtue of its entirely glossy, purple-black plumage. The female is duller and has a gray collar and underparts. In flight the wings appear short and triangular, while the tail is broad and forked. Martins can be attracted to communal nest boxes in open areas. One of the most beneficial of all species, it consumes large quantities of flying insects, particularly mosquitos, which are taken on the wing. A common migrant and breeding species in southern Florida.

Male (above); female (left)

Tree Swallow *Tachycineta bicolor* 13–15cm

Its upperparts are dark greenish-blue, the dark crown extends to the eye and the underparts are bright white. The tail is notched. Frequents wet prairie areas, especially in the Everglades, and nests in woods, fence posts, under eaves and in nest boxes. Feeds on flying insects, but also takes wax myrtle berries. Commonly observed over water, where individuals flutter near the surface to drink. Appears hunch-backed in flight and lowers its wingtips when gliding. An abundant migrant and winter resident in southern Florida.

Male (top); female (bottom)

Northern Rough-winged Swallow *Stelgidopteryx serripennis* 12–15cm

Similar to the Bank Swallow but lacks that species' brown breast-band. The upperparts are brown, the throat gray-brown, the lower breast and abdomen are white and the tail is notched. Frequently observed along riverbanks, road cuts, gravel pits, dams and bridges. In flight the wingtips appear to be restrained after the downstroke, but its movements are slow and fluid. Usually seen in pairs or small groups. An uncommon migrant and winter resident to southern Florida; it nests in northern Florida.

Cave Swallow *Petrochelidon fulva* 12–15cm

R. & N. Bowers/VIREO

A compact, attractively plumaged aerial feeder. The back is blue-black with narrow white striping and the rump is bright buff. It has a blue-black cap and a bright chestnut forehead. The tail is squared off. Frequents open country, wetlands and arid scrub in the vicinity of water where it feeds on flying insects. Its usual breeding sites are in culverts and under bridges. Once a rare migrant to southern Florida, in recent years the West Indian form, which may be a separate species from those in Texas and Central America, has become a summer resident and common nesting species.

Barn Swallow *Hirundo rustica* 15–19cm

Widespread and familiar. Adults are shiny dark blue above and tan below, with a reddish-brown throat. The tail is deeply forked. It is a swift flyer and very acrobatic as it skim-feeds near the ground in open areas. Exclusively takes insects, which it snatches while in flight. It has a tendency to frequent areas where cattle, other mammals and birds stir insects into flight. Individuals often skim still water to glean insects from the surface. An abundant summer migrant in southern Florida. Observed during all months of the year.

Tufted Titmouse *Baeolophus bicolor* 15cm

This is a conspicuous species at backyard feeders, and a common resident in the north-east section of southern Florida. The upperparts are mouse-gray, it has a crest and a black forehead, the flanks are buffy, while the rest of the underparts are pale gray. Acrobatic, it hangs upside-down while foraging at all levels in evergreen and deciduous woodlands. Feeds primarily on insects, mainly caterpillars, but also on berries, seeds and acorns. Its *peter-peter* song is a familiar sound in suburban areas.

Brown-headed Nuthatch *Sitta pusilla* 11cm

This species, the only nuthatch to occur in Florida, is gray above. There is a white nape spot and the crown is brown, and it has a black eye-stripe. The throat is white, the flanks and belly pale gray, and the breast is buff. Usually observed creeping upside-down along trunks and main branches of Loblolly pines, searching the crevices for insects, and prying seeds from pine cones. A rare resident of mature evergreen forests, which has declined owing to logging. In southern Florida it is now found only in the vicinity of the Big Cypress Swamp.

Carolina Wren *Thryothorus ludovicianus* 14cm

With its bright buff underparts, white eyebrow and entirely rufous upperparts, including the crown, this is a distinctive species. The tail is long and regularly cocked, while the bill is long and slightly decurved. Usually observed in dense vegetation near the ground. An active, curious species and frequent singer. Its loud song has been likened to *tea-kettle, tea-kettle*. Feeds on insects, berries and seeds. This familiar resident is a common denizen of all wooded areas of southern Florida including hammocks and cypress swamps.

House Wren *Troglodytes aedon* 11–13cm

Typically appears as a gray-brown, unstreaked bird, with a pale supercilium and barred tail; there is some black barring on the upperparts. The underparts are pale gray. Frequents thickets and the understory, especially of hammocks and palmetto, and is also common in urban areas. Feeds mainly on insects. Its song is a melodious cascade of bubbling notes. In southern Florida this species is a common migrant and winter visitor.

Sedge Wren *Cistothorus platensis* 10–12cm

Uniform brown in color, with barred wing-coverts, it has a plain face, streaked crown and buffy-orange flanks. There are some faint white stripes on the back and the bill is short. Its preferred habitat is wet fields and the damp edges of marshes and bogs. Frequents both fresh and brackish water. Very shy, when flushed it flies only a short distance before dropping again into dense vegetation. Feeds on insects and spiders. Formerly known as the Short-billed Marsh Wren, this species is an uncommon winter visitor to marshes and prairies of southern Florida.

Marsh Wren *Cistothorus palustris* 10–14cm

Dark brown upperparts are darkest on the crown. The supercilium is white and the mantle striped with white. The flanks are orange, the chin and belly white, while the bill is long, slender and slightly decurved. Not as shy as the Sedge Wren, but secretive nonetheless. Its song is a raspy unmusical buzz. Frequents tall, dense, emergent vegetation especially cattails in fresh and brackish water marshes, where it feeds on insects and spiders. Constructs a football-shaped nest in reeds over water. Formerly known as the Long-billed Marsh Wren, it is a common migrant and winter visitor to southern Florida.

Red-whiskered Bulbul *Pycnonotus jocosus* 18cm

The black cap and very large crest are unlike any other species in our region. There is a bright red ear patch; the upperparts are brown and the underparts white. The undertail-coverts are also red and the tail is long with white spots at its end. Frequents ornamental and exotic vegetation, feeding on fruit, flowers, insects and small lizards. Gregarious, occupying communal roosts during the non-breeding season. Commonly seen on utility wires, this bulbul became established in 1960 after a small group escaped from an aviary. It is now a common breeder in southern Florida. Originally an Afro/Asian species.

Ruby-crowned Kinglet *Regulus calendula* 10cm

A tiny bird, with greenish-olive upperparts, a broken white eye-ring and two white wingbars. The male has a bright red crown patch that is normally concealed. Habitually flicks its wings as it chases insects, which it feeds on almost exclusively. Constantly active. Forages in trees and shrubs at all levels and is especially fond of leaf clumps. Found in hammocks and woodlands, it is relatively tame and readily responds to 'pishing'. Calls frequently, a variety of thin, high-pitched notes. An uncommon migrant and winter visitor to southern Florida.

Blue-gray Gnatcatcher *Polioptila caerulea* 11cm

This small active species has gray upperparts and a long thin black tail, with white outer feathers, that it frequently cocks. The underparts, as well as its conspicuous eye-ring, are white. Warbler-sized, its pattern is reminiscent of a mockingbird. Frequents hammocks, cypress swamps and pine woodlands. Very tame, its diet principally consists of insects gleaned from the undergrowth. Very responsive to 'pishing'. The call is a high-pitched buzz. A common migrant and winter visitor to southern Florida.

Male (top); female (bottom)

Eastern Bluebird *Sialia sialis* 18cm

Male

Female

Unmistakable, the male's upperparts are bright blue; the throat, breast and flanks are chestnut, and the belly and undertail-coverts are white. The coloration of the female is paler. It frequents open woodlands and agricultural land. Feeds primarily on insects, taking wild fruit and berries during cooler periods of the year. An uncommon resident species of southern Florida. Formerly more common in pine forests, but modern logging practices have felled the many of the hollow trees upon which it depends for nest sites.

Veery *Catharus fuscescens* 16–19cm

Small, shy thrush that spends most time near the ground. Its upperparts are uniform tawny-brown including the tail, while the breast is buffy with very pale brown spots, and the rest of the underparts are white. There is a very pale eye-ring. The species feeds on insects and is usually skulking. This uncommon thrush is silent on migration through Florida to and from its wintering grounds in Central and northern South America.

Gray-cheeked Thrush *Catharus minimus* 18–20cm

Poorly marked, the Gray-cheeked Thrush has an indistinct and incomplete white eye-ring and gray cheeks. There is heavy spotting on the gray breast (not usually buffy as in Swainson's Thrush), while the rest of the underparts are white, becoming olive-gray on the flanks. The upperparts are olive-brown. Largely silent while on passage to its breeding grounds in coniferous and mixed forests at boreal latitudes, it frequents the forest floor where it forages in the leaf litter for insects. An uncommon migrant through southern Florida.

Swainson's Thrush *Catharus ustulatus* 18cm

Formerly known as the Olive-backed Thrush, this species is similar to the Gray-cheeked Thrush, but can be easily distinguished, given a clear view, by its prominent buffy eye-ring. The upperparts are olive-brown and the breast is buffy with heavy spotting. The sides of the head and throat are also buff, and the rest of the underparts are white, becoming olive-gray on the flanks. Feeds on the forest floor, sifting through the debris for insects and other animal material. It is an uncommon spring and fall migrant in southern Florida.

Hermit Thrush *Catharus guttatus* 18cm

Another subtly marked, basically small brown thrush. The upperparts are brown, diagnostically contrasting with reddish-brown tail. The underparts are whitish-buff, darkly spotted on the breast. There is a thin, complete, usually whitish eye-ring. Frequents hammocks and wooded wet areas. Feeds on insects, fruit and berries gleaned from the forest floor. Habitually cocks and lowers its tail while flicking its wings. An uncommon migrant and winter resident in southern Florida.

Wood Thrush *Hylocichla mustelina* 20cm

Larger than the other predominantly brown thrushes. Bright cinnamon head, with paler back and tail that grade into olive-brown. Conspicuous white eye-ring. The bright white underparts are completely covered with very large black spots. This thrush is usually encountered in wooded areas near water where it feeds on a variety of insects and berries. It is not shy and commonly occurs in urban areas. When alarmed it raises the feathers of the crown in the suggestion of a crest. Very small numbers winter in southern Florida, but this species is principally a migrant.

American Robin *Turdus migratorius* 23–28cm

Adult

Immature

One of the most familiar of North American birds. Strikingly plumaged: the head is black, with white spectacles and a yellow bill, the back is gray and the underparts, with the exception of the white undertail-coverts, bright orange-rufous. Larger than other North American thrushes. Feeds on earthworms and insects, and is fond of the introduced exotic Brazilian pepper berries. This common migrant and winter visitor to southern Florida is observed in a variety of habitats including hammocks, farmland, pines, on coasts and even in urban areas, usually in flocks.

Gray Catbird *Dumetella carolinensis* 23cm

This almost entirely gray bird has a black cap, chestnut undertail-coverts and a long tail, which it cocks frequently as it forages on or near the ground. Frequents dense undergrowth, but has benefited from the recent human population explosion, being able to survive well in residential areas. Its cat-like *meow* is distinctive, and like mockingbirds it is a fine mimic of other bird's calls. Feeds on small fruits, berries and insects. Common and widespread throughout southern Florida.

Northern Mockingbird *Mimus polyglottos* 23–28cm

Gray above and grayish-white below. The wings and tail are distinctly marked with white, which is usually conspicuous in flight. The tail is frequently cocked. It is relatively tame. Feeds on insects and fruits, as well as garbage. A remarkable mimic that sometimes sings at night. A common resident of southern Florida, it frequents open country, urban areas, semi-arid scrub, mangroves, hammocks, pines and citrus groves, and persists well around habitation. In the Bahamas and Cuba it appears to be displacing the Bahama Mockingbird. An aggressive species.

Bahama Mockingbird *Mimus gundlachii* 28cm

Larger than the Northern Mockingbird: the upperparts are brownish-gray, finely streaked with dusky on the head and back. There are two indistinct wing-bars, the underparts are whitish, also streaked dusky, and the tail is tipped with white. Like the Northern Mockingbird it is an excellent mimic. It feeds on insects and fruits. Principally occurs in the Bahamas and, more marginally, in parts of the Greater Antilles, where it prefers taller, denser vegetation than the Northern Mockingbird. Regularly recorded on the south-east coast of Florida, where it is almost annual.

Brown Thrasher *Toxostoma rufum* 29cm

The upperparts are bright rufous, including the long tail, while the underparts are buffy-white streaked heavily with black. The decurved bill is black and the eyes are yellow. Its primary habitat is dense thickets within woodlands. It nests in natural habitat, but can adapt to urban gardens and backyards. Feeds on berries, insects and seeds that it locates at or near ground level. It sings loudly from an exposed perch, but unlike its close relatives, the mockingbirds, it only rarely imitates other birds. The Brown Thrasher is a common resident of southern Florida.

Common Myna *Acridotheres tristis* 25–26cm

Mike Lane; Aquila

Introduced to southern Florida, probably through escaped cage birds. It now breeds here. Brown with a glossy black head, the bill, a bare patch around the eye and the legs are yellow, while the wing patch, terminal tail band and undertail-coverts are white. In flight the brown body contrasts with the black rounded wings and black tail; the white wing patches and terminal tail band are obvious. Frequents urban areas including parks, backyards and other open land. Cursorial, it principally feeds on insects, fallen fruit, seeds, grain and nectar, but also is a scavenger and occurs around garbage.

Hill Myna *Gracula religiosa* 23–30cm

Chris Gomersall; Windrush Photos

This myna is large and almost entirely black, glossed purple. There is a white wing patch. The bill is orange, while the legs yellow and there are yellow wattles below the eye and on the nape. Mainly frugivorous, flocks assemble in favoured fruiting trees. Strictly arboreal, it is a cavity nester. Monogamous, it pairs for life. Sidles sideways along branches and hops on the ground. Known in the pet trade as the 'talking myna', this common cage bird has become introduced through escapees and is now a breeding resident in south-east Florida.

Golden-winged Warbler *Vermivora chrysoptera* 11–12cm

Males have a black mask and throat, the back and wings are blue-gray, and there is a large yellow wing patch that consists of two broad wingbars. The underparts are white and the crown is bright yellow. The female is duller in comparison. This species occurs near hammocks and in brushy areas. Similar in habits to the Blue-winged Warbler, the two hybridize freely, producing at least two well-marked intergrades. This species is declining as the Blue-winged Warbler is invading its former range. Feeds almost entirely on small insects. This uncommon species migrates along both coasts of Florida.

Nashville Warbler *Vermivora ruficapilla* 11–13cm

Short-tailed warbler. Gray head with a full white eye-ring, the rest of the upperparts, including the tail, are yellow-green. The throat and breast to the undertail-coverts are yellow, with white thighs. This active warbler is usually observed moving through the lower branches of broadleaved trees. Forages at all levels of second-growth woodland; it is entirely insectivorous. An uncommon migrant through coastal Florida.

Orange-crowned Warbler *Vermivora celata* 13cm

The upperparts are olive-green; the underparts are yellow-olive. There is faint streaking on the breast-sides and a pale supercilium that is not always present. The bill is thin and very slightly decurved, but the orange crown patch is usually concealed. There are few distinctive field marks. Frequents scrubby oak woodland and other brushy habitats. Appears to be largely coastal during its northbound migration. Feeds high in the canopy during spring, and in brushy or weedy areas in the fall. Feeds on insects, berries and small fruits. An uncommon migrant and winter visitor to southern Florida.

Northern Parula *Parula americana* 11cm

Attractive and easily recognized species. The upperparts are blue-gray, except for a triangular yellow patch in the center of the back, and there is an incomplete white eye-ring. The underparts are bright yellow on the breast and white on the abdomen. There is a band of black and chestnut across the chest. Joins mixed-species flocks in winter. It frequents cypress swamps and hammocks. Feeds on insects, taken while gleaning leaves within the canopy. A common migrant and winter visitor to southern Florida; a few pairs nest.

Yellow Warbler *Dendroica petechia* 12cm

Male

Female

One of the most abundant and well-known wood warblers, its plumage is variable, but is generally entirely yellow or gold above and below. Males can have chestnut streaking on the breast, and some races have variable amounts of chestnut on the head. It maintains feeding territories during the winter. Nests are parasitized by the Shiny Cowbird, a recent southern Florida invader. The bright West Indian race is a common resident of mangroves in Florida, while the North American race is a passage migrant and rare winter visitor.

Chestnut-sided Warbler *Dendroica pensylvanica* 11–14cm

Male

Autumn plumage

An unusually striking bird. The male's crown is bright yellow, contrasting with the black-and-white facial pattern, while true to its name, the flanks and sides are chestnut, there are a pair of bright yellow wingbars and rest of the underparts are white. The female's colors are more subdued. Frequents thickets and second growth and is wholly insectivorous. This was a rare species in the 1800s, but changing land-uses have enabled it to increase dramatically and it has consequently become more common on migration through southern Florida in recent years.

Magnolia Warbler *Dendroica magnolia* 12cm

Male

Female

Breeding males are highly distinctive. The crown is gray; the back is black and the rump yellow. The underparts are bright yellow, heavily streaked with black and the undertail-coverts are white. There are two bright white wingbars and a supercilium. The sides of the head are black. White tail patches are apparent in flight. Females are much duller. Frequents hammocks and brushy wet areas, as well as weedy fields in the fall. Habitually flits from perch to perch, constantly spreading its tail. Uncommon migrant and rare winter visitor to southern Florida.

Cape May Warbler *Dendroica tigrina* 12cm

Adult

Immature

Identified by its chestnut cheeks, white wingbar, black-streaked yellowish underparts and a yellow to greenish rump. Female underparts are not as heavily streaked and they also lack the cheek patch and wingbar. In all plumages there is a trace of yellow at the sides of the neck. Feeds on nectar and fruit, as well as insects, during migration and on the wintering grounds. The thin, slightly decurved bill is well suited to this purpose. Frequently observed in gardens and hammocks. Numbers vary from year to year. A common migrant in southern Florida; small numbers overwinter.

Black-throated Blue Warbler *Dendroica caerulescens* 13cm

A distinctive and boldly patterned warbler. Males in breeding plumage are blue above and white below, with a black face, throat and flanks. A bold white spot, which is sometimes concealed, is present at the base of the primaries in both sexes. It assists in the identification of the green and yellow females, which also have a white eyebrow (not shown by the male). Forages in low vegetation, feeding on nectar, fruit and insects. Males generally forage higher in trees than females. A common passage migrant in southern Florida. A few also winter there, but most of the population winters in the West Indies.

Male (top); female (bottom)

Yellow-rumped Warbler *Dendroica coronata* 14cm

The upperparts are blue-gray; the underparts are white, streaked with black. In all plumages the bright yellow rump, crown patch and spot on the side of the breast is indicative of this species. Eastern populations have a white throat. Travels in small groups, feeding on insects and berries. Adapts to a variety of food sources, its foraging methods are the most diverse of any warbler. The *check* call note is distinctive. Formerly known as the Myrtle Warbler, this species is common to abundant in southern Florida as both a passage migrant and winter visitor.

Black-throated Green Warbler *Dendroica virens* 12cm

In all plumages, this species has grayish-yellow cheeks surrounded with yellow. Adults have a black throat and flank stripes, and two conspicuous, white wingbars. The mantle is bright olive and the rest of the wings black. Rear underparts white. Feeds primarily on insects, especially caterpillars; it is an arboreal species that prefers forest canopies, especially those of pine and Cypress. Moves quickly through the branches, gleaning its prey from the leaves. An uncommon migrant and occasional winter visitor to southern Florida.

Blackburnian Warbler *Dendroica fusca* 12–14cm

This is a spectacular warbler. The male has a flaming orange crown patch, eyebrow and throat; the back is black streaked with white, and there is a large white patch in the otherwise black wing. The underparts are white, with fine black streaking on the flanks. In female plumage the orange is replaced by yellow. Normally a treetop denizen, it frequents lower levels during migration. Diet is primarily insectivorous. Cocks its tail and droops its wings while foraging. An uncommon passage migrant in southern Florida.

Yellow-throated Warbler *Dendroica dominica* 13cm

This warbler has a bright yellow throat and a striking black-and-white face pattern. The underparts are white and there is heavy black streaking on the flanks. It has two distinct, white wingbars. The majority of the upperparts are dark blue-gray. Comparatively long billed. A canopy species, it commonly creeps along trunks and major branches. It prefers pines, hammocks, mangroves and Cypress edges, and is one of the earliest fall migrants to arrive, usually appearing in late summer. A common migrant and winter visitor to southern Florida; a few remain the summer but do not breed.

Pine Warbler *Dendroica pinus* 13–15cm

Large and heavy-billed warbler, with two whitish wingbars. The upperparts are olive and unstreaked; underparts are yellow, fading to whitish on the belly, and there is faint variable streaking on the sides of breast and flanks. As befits its name, it is a common resident of the pine forests of southern Florida, where it feeds on insects, spiders, seeds and berries, although it is also found in other habitats in the area. It has a creeping locomotion as it feeds, at all levels from the ground to the canopy. Its winter range includes most of the breeding area. It can sometimes be heard singing in the winter.

Adult (top); immature (bottom)

Prairie Warbler *Dendroica discolor* 12cm

This tail-bobbing warbler is pale olive-gray above and yellow below. Variable amounts of, usually prominent, black streaking appear on the flanks; there is a black crescent below the eye and, in most, a distinctive yellow-and-black facial pattern. Two pale yellowish wing-bars are also present. Frequents mangroves and other brushy areas in coastal areas, and is generally fond of second growth. Feeds on insects and spiders, occasionally nectar. A common migrant and winter visitor to southern Florida; there is a breeding population.

Palm Warbler *Dendroica palmarum* 13cm

Tail-wagging species that is yellow or whitish below, but has yellow undertail-coverts in all plumages. It also has a brownish back, white eye-stripe and olive rump. There are two forms of this species that are separable in the field: the western form has largely white underparts, those of the eastern form are yellow. It is usually observed on the ground or in shrubby, low vegetation, and frequents pines, hammocks, wet edges, second growth and urban areas. Despite its name, it has no affinity to palms. The rarer eastern form appears to prefer more wooded habitats. This warbler is a common migrant and winter visitor to Florida.

Bay-breasted Warbler *Dendroica castanea* 12–15cm

Male breeding plumage

Autumn plumage

In spring the male has a chestnut crown patch, throat and flanks, a buffy neck patch and white wingbars. The upperparts are greenish-gray streaked with black, while the rest of the underparts are white. The female is similar, but paler. In late summer and fall, the striking spring finery is replaced by a much drabber plumage, which makes the species difficult to identify. At this season the upperparts are gray-green and streaked, the buff underparts are unstreaked, although there is a trace of reddish on the flanks, while the undertail-coverts are buff and the legs dark. A slow deliberate feeder, on insects, from mid-level to the canopy. This warbler is an uncommon migrant in coastal southern Florida.

Blackpoll Warbler *Dendroica striata* 13cm

The male has a black cap and white cheeks; the upperparts are streaked olive-gray. There are two white wingbars and the underparts are white, with black-streaked flanks. The female is similarly patterned, but has a greenish cap and eye-stripe; its underparts are streaked yellowish-green. In fall this species is distinguished from the similar Bay-breasted Warbler by its yellow legs, streaked underparts and usually white undertail-coverts. Frequents the canopy of hardwood forests, where it travels slowly, gleaning leaves for insects. A common spring migrant in Florida, but rare in fall. Song is very high pitched.

Male (top); fall (bottom)

Black-and-white Warbler *Mniotilta varia* 13–14cm

This familiar species' distinctive habit of creeping up and down tree trunks and branches is unique among wood warblers. To accomplish this it has developed strong legs and a long bill. This adaptation may permit it to leave its wintering grounds earlier and return later than other species. Its entirely black-and-white coloration and striped crown are diagnostic. Frequents woodlands of all sorts, as well as hammocks. Feeds on insects and spiders, many of which are gleaned from bark crevices. Common as a migrant and winter resident in southern Florida.

American Redstart *Setophaga ruticilla* 13cm

Male (top); female (bottom)

A very active warbler that constantly displays its colors by fanning the tail and flicking the wings. The flashy male has black upperparts, with orange patches at the base of the flight feathers and rectrices; the female is greenish-gray, the orange being replaced by yellow. Frequents mangroves and hammocks. It feeds on insects, berries and seeds, usually on the forest floor to mid-level, but also occasionally snatches insects in the air. The redstart is a common migrant and infrequent winter resident in southern Florida.

Prothonotary Warbler *Protonotaria citrea* 13cm

Dennis Bozzay

This uncommon migrant occasionally nests in tree cavities within the cypress swamps of southern Florida. The head and breast of the male is bright orange-yellow, the wings are blue-gray, the back olive and the tail darker with prominent white patches. In comparison, the female is paler yellow. This species is most often observed foraging in wet woodlands and mangroves, usually at ground level. Primarily insectivorous. It is a birder's favorite owing to its brilliant plumage and loud ringing song. Sometimes called the Golden Swamp Warbler.

Worm-eating Warbler *Helmitheros vermivorus* 14cm

This warbler has greenish-gray upperparts, a brown crown with a black stripe through it and another through the eye. The underparts are white, grading to buff. Frequents forested areas with heavy leaf litter, dead leaf clusters and heavy vines, usually in the lowlands and foothills. Feeds almost entirely on insects and spiders that it takes from on, or near, the ground. It is solitary and territorial on its wintering grounds. When walking on the ground it maintains its tail in a cocked position. An uncommon migrant; very small numbers may remain to winter in southern Florida.

Swainson's Warbler *Limnothlypis swainsonii* 13cm

Kevin T. Karlson

This skulking species is a ground dweller in forested areas, usually near water. Its plumage is olive-brown above and yellow-white below. It has a reddish-brown cap and a light superciliary, and the bill is large and pointed. Usually occurs in hammocks or wooded swamps, but on migration is more widespread, provided there are dense clusters of dead leaves. It walks rather than hops, noisily picking and overturning leaves in its search for insects. Remains motionless for long periods when alarmed. An uncommon to rare migrant in southern Florida.

Ovenbird *Seiurus aurocapillus* 15cm

Adult

Immature

The thrush-like Ovenbird is, belying its appearance, actually a wood warbler and is easily identified by its manner of walking while feeding on the forest floor: it calls and bobs its head, cocking its tail frequently. The upperparts are brown, with an orange crown, bordered by black stripes, and a bright white eye-ring. Underparts white, heavily spotted with black. Feeds on insects, spiders, earthworms and snails gleaned from the leaf litter on the forest floor. Avoids wet lowland forests, but sometimes occurs in mangroves. A common migrant and winter visitor to southern Florida.

Northern Waterthrush *Seiurus noveboracensis* 13–15cm

Dark brown above and pale buff below, with dark brown streaks. There is a prominent buffy eyebrow, which narrows behind the eye and fine streaking on the throat. The eye-stripe of the Louisiana Waterthrush extends behind the eye, and is white or bicolored, while that species' throat is clear and unstreaked. Northern Waterthrush has a longer tail. Feeds at the edges of watercourses, teetering along as it snatches insects and their larvae. Frequents mangroves and other swampy areas, especially along the coast. Common migrant and winter resident.

Louisiana Waterthrush *Seiurus motacilla* 15cm

Grayish-brown above and white below with dark brown streaks. The white eye-stripe, which broadens behind the eye, and the pure white throat, serve to separate this species from the very similar Northern Waterthrush. Like the preceding species, it bobs and teeters along water edges; it prefers fast-flowing streams with forested edges, and is rarer along the coast. The bobbing motion of this species is slower and more deliberate, and it is shyer than the Northern Waterthrush. Feeds on small mollusks, earthworms, crustaceans and even small fish. This uncommon migrant occasionally summers in southern Florida, but does not breed.

Common Yellowthroat *Geothlypis trichas* 11–14cm

Male

Female

One of the most familiar and widespread of warblers, the male's black mask with an upper rim of white is wholly distinctive of this species. The female has no mask, but both sexes sport a bright yellow throat and breast, and a whitish belly. The underparts, from the mantle to the tail, are olive-green. Common in wet brushy fields, marshes and dense vegetation, where it feeds on insects and other invertebrates. Its distinctive song *witchity-witchity-witch* is sung from a prominent perch. Perches with its tail cocked. A common Florida resident.

Hooded Warbler *Wilsonia citrina* 12–15cm

Male (top); female (bottom)

Easily distinguished by its bright yellow face and black hood and throat, the underparts are bright yellow and its upperparts are olive-green. The female is duller and has only a trace of the hood. Both sexes have extensive white in the tail. Frequents the dense understory, especially of hammocks and swamps, where it flits actively, chasing insects in the manner of a flycatcher. Rapidly opens and closes the tail laterally, flaring the white outer feathers. Its presence is an important indicator of forest quality. This species is an uncommon migrant in southern Florida.

Yellow-breasted Chat *Icteria virens* 19cm

The largest wood warbler. Along with its size, the white spectacles, heavy black bill, bright yellow chin, throat and breast, and olive-green underparts serve to distinguish this species. Occurs in brushy areas within dense vegetation. Solitary, it is usually rather noisy as it forages for insects and berries. A very curious species that readily responds to 'pishing'. It is an uncommon migrant to southern Florida; few overwinter.

Bananaquit *Coereba flaveola* 10–12cm

The white eye-stripe is diagnostic in most plumages. Secondary field marks are the white throat and yellow underparts, white wing patch and a sharp, decurved bill; the tail is black with white spots. Occurs in hammocks, gardens and thickets. Frequents flowering shrubs where it feeds on nectar, fruit, spiders and insects. When feeding it punctures the base of the flower to obtain the nectar. Acrobatic, sociable and noisy. One of the most abundant species in the West Indies, Central and South America, it is a casual visitor to southern Florida in spring and fall. Only the Bahama race occurs in Florida.

Summer Tanager *Piranga rubra* 18–20cm

Male

Female

Adult males are entirely rosy-red year-round. Females and immatures have a yellow face and underparts, and a dark olive back and wings. The bill is large and pale in all plumages. Commonly observed in the woodland canopy, where it forages for insects by gleaning the leaves in a leisurely fashion; it is particularly fond of bees and wasps. An uncommon migrant and summer visitor to southern Florida, which formerly nested.

Scarlet Tanager *Piranga olivacea* 16–19cm

Male

Female

A. & E. Morris; Windrush Photos

The male Scarlet Tanager is one of the most striking American birds. Its plumage, with the exception of the black wings and tail, is entirely brilliant red. The female has dark wings, greenish upperparts and yellow underparts. In southern Florida it is a regular spring and fall migrant, where it is more common in the west. It frequents the canopy of deciduous trees as it searches for small fruits and insects. This species moves about slow and deliberately, and sometimes goes to the ground for insects.

Western Spindalis *Spindalis zena* 14cm

Male

Female

Formerly known as Stripe-headed Tanager, males are black-headed, with a striking white supercilium and malar stripe, the breast is yellow the rest of the underparts are white, and there are bright chestnut patches on the nape and rump. Females are significantly duller, being grayish-olive above and whitish below. Occurs in hammocks, brush, open woodlands and second growth, always near the ocean. Tends to perch high and unobtrusively in the canopy when not feeding on berries, tender leaves and other plant material. Insects are apparently not an important dietary constituent. A rare vagrant to southern Florida that has been recorded in all seasons except summer.

Northern Cardinal *Cardinalis cardinalis* 19–23cm

Male (top); female (above)

A familiar, well-loved garden and feeder bird. The male is unmistakable with its largely red plumage and crest, black mask and bib and deep, large reddish-orange bill. The female is similar in shape, but the bill is pink, upperparts are gray-olive and wings and tail reddish. Frequents thickets, palmetto areas with vines. Feeds on large seeds and berries, usually in pairs or family groups. The song is a loud *cheer-cheer-cheer*, delivered from a perch, with the tail cocked and the crest raised. A common resident of southern Florida.

Rose-breasted Grosbeak *Pheucticus ludovicianus* 18–21cm

Arthur Morris; Windrush Photos

Unmistakable. The spring male has a solid black head and upperparts, the black wings have striking white patches, while the breast is red and the rest of the underparts white. The female has a pale white eyebrow, a dark brown crown with a median stripe, the breast and flanks are streaked and it has yellow wing linings. Both sexes have a pale-colored heavy bill. Frequents dense woodlands where it forages for fruits and insects. Small numbers winter in the West Indies and occur as uncommon migrants in southern Florida.

Male (top); female (bottom)

Indigo Bunting *Passerina cyanea* 14cm

Kevin T. Karlson

Readily identifiable in all plumages. The spring male is entirely indigo-blue, as its name implies, and the bill is short and black; the upper mandible is curved. The female and winter male are brown overall, with faint breast streaks and indistinct wingbars. Immature and moulting males usually display some patchy blue on both the upper and underparts. Occurs in prairies, brushy areas and woodland edges. Feeds on insects, seeds and grains. Considerably smaller than the Blue Grosbeak. A common migrant and winter resident to southern Florida.

Painted Bunting *Passerina ciris* 13–14cm

This rainbow-colored, rather dumpy species is also known as the 'Nonpareil'. The male has a blue head, red orbital ring, green back, red underparts and rump, and a yellow-green back. Less distinctive, the female is dark green above and paler, yellow-green below; both sexes have short conical bills. Frequents hammocks, open brushy areas and even backyards, where it is a common sight at feeders and birdbaths. Its standard food is small seeds, insects and berries. A common migrant and winter visitor to southern Florida.

A. & E. Morris; Windrush Photos

Wayne Lankinen; Aquila

Male (top); female (bottom)

Eastern Towhee *Pipilo erythrophthalmus* 18–22cm

Arthur Morris; Windrush Photos

Male

A. & E. Morris; Windrush Photos

Female

Another readily identifiable species. Males have a black head, breast and upperparts, while the breast-sides and flanks are rufous, and the rest of the underparts white. The head, breast and upperparts are brown in females. It occurs in brushy edges, pine forests and hammocks. Cursorial, it hops along, raking through the litter foraging for insects, seeds, berries and other material. Common around the base of backyard feeders. Its song, *drink your tea*, is distinctive. This towhee is a common and widespread southern Florida resident.

Bachman's Sparrow *Aimophila aestivalis* 12–16cm

Formerly known as the Pine Woods Sparrow, this is an uncommon and declining resident of the pine woods in southern Florida. Its upperparts and eye-stripe are reddish-brown, the underparts are buffy and unstreaked, the bill is large and the tail rounded. Highly secretive, it occurs in pines and palmetto scrub. When alarmed, Bachman's Sparrow runs along the ground and, if flushed, flies low and far, usually freezing on a low perch. Often shares the same habitat as the Red-cockaded Woodpecker. Feeds on insects and seeds.

Chipping Sparrow *Spizella passerina* 12–14cm

Kevin T. Karlson

Distinguished by its bright rufous cap, white supercilium, and black eye-stripe and lores. Cheeks, underparts and rump are pale gray; the back is brown, streaked with black. The tail is long and notched. Non-breeders have paler heads. Small flocks are usually observed at ground level, while foraging near woodland edges, yards and weedy fields. More arboreal than most species of sparrow. Feeds mainly on seeds, but some insects are also taken. An uncommon migrant and winter visitor to the coastal plain of southern Florida.

Savannah Sparrow *Passerculus sandwichensis* 11–15cm

Although rather unassuming in plumage, the yellow superciliary, along with its streaked upperparts and underparts are diagnostic. The tail is short and notched, and the legs pink. Occurs in wet grasslands throughout, including coastal areas. It feeds on the ground taking insects and seeds. Frequents dense cover and, when flushed, it flies a short distance before dropping back into the vegetation; the flight is strong and direct. A common passage migrant and winter resident in southern Florida.

Saltmarsh Sharp-tailed Sparrow *Ammodramus caudacutus* 11–13cm

Inhabits drier areas of marshes, where it skulks, mouse-like, within dense grasses, often running rather than flying. The crown, nape and face are gray, bordered with orange-buff. The upperparts are dark brown and the underparts are white with dark streaks on an ochre breast. The tail is short and pointed. Recently been separated from the inland form, Nelson's Sharp-tailed Sparrow (*A. nelsoni*), shown here, and now considered a separate species. In flight it holds its head raised and tail lowered. Feeds almost entirely on animal material, especially insects. This species is an uncommon passage migrant and winter visitor to southern Florida. Nelson's Sharp-tailed Sparrow has a paler facial pattern, and no breast streaking; it is more common along the Gulf Coast.

Seaside Sparrow *Ammodramus maritimus* 12–15cm

Kevin T. Karlson

Easily distinguished by the yellow patch in front of the eye, pale moustachial stripe and a dark malar; the upperparts are olive-gray and there are dusky streaks on the underparts. The tail is short and pointed. In southern Florida this species consists of two races, which were formerly considered separate species. The uncommon Gulf coast race is a migrant and winter visitor, whereas the olive-green Cape Sable race is resident in the Everglades. Both forms occur in saltmarshes, where they feed on insects and other animal material.

Swamp Sparrow *Melospiza georgiana* 12–15cm

An unobtrusive sparrow of wet areas. The upperparts are rufous-brown, the crown and wings chestnut, and the face and breast are gray. Underparts white, with a broad gray breast-band, which bleeds onto the flanks. There is some fine streaking on the breast and flanks. Its bill is more slender than most sparrows, indicative of its more insectivorous diet. Frequents wet brush, usually near the ground; wades in shallow water where it picks insects and seeds from the surface. Common as a migrant and winter visitor in Florida.

Red-winged Blackbird *Agelaius phoeniceus* 19–23cm

Male

Female

Abundant resident of Florida, the male Red-winged Blackbird is a familiar, noisy and unmistakable species. Shiny black, with a large scarlet shoulder patch, it contrasts sharply with the female, which has dark brown upperparts, a pale eyebrow and heavily streaked underparts. Occurs year-round in large flocks within freshwater marshes, but also visits cultivated land and pastures, where it feeds on seeds, fruit, insects and even small invertebrates. Large flocks can cause considerable damage to rice and other grain crops.

Eastern Meadowlark *Sturnella magna* 20–26cm

This meadowlark is a familiar species. The upperparts are streaked brown and the underparts are bright yellow; there is a black V on the breast, and obvious white outer tail feathers. Perches on fences and wires, and other low perches, at all elevations where suitable habitat exists. It is a ground feeder, mainly taking insects and other animal material. Frequents open areas, cultivated land, grasslands and marshes. Has a distinctive wobble as it walks along the ground. Common resident of Florida.

Boat-tailed Grackle *Quiscalus major* 33–41cm

Male (top); female (bottom)

Unmistakable, the largest blackbird, the male is iridescent blue-black, with a dark brown eye and long keeled tail. Females have the upperparts glossy dark brown, while the breast is tawny-brown, but share the distinctive tail shape and eye color. Occurs in marshes both coastal and inland, where it nests colonially in low shrubbery. A cursorial species, it wades into shallow water to feed on small fish, crayfish, snails, insects and other animal material. It is an opportunist and will forage through garbage in urban areas. Frequents parks and yards. An abundant resident in Florida.

Common Grackle *Quiscalus quiscula* 28–33cm

Bronze phase

Purple phase

It is the purple rather than the bronze phase that occurs in Florida. Males are entirely black with a purple iridescence to the head, neck and breast. The eyes are yellow. Though considerably smaller than the Boat-tailed Grackle, the long tail is also spatulate in shape. Juveniles are brown. Occurs in all habitats, from residential areas in cities to farmland. Frequents cypress swamps, marshes and hammocks. Catholic diet includes lizards, insects, snails, seeds and berries. This blackbird is abundant in southern Florida.

Brown-headed Cowbird *Molothrus ater* 19cm

A successful brood parasite that lays its eggs in nests of other species. The male has a brown hood, short conical bill, glossy black body and short tail; the female has gray-brown plumage that is paler below. Distinguished from other blackbirds by its short wings and tail. Feeds on insects and seeds, with other blackbirds, usually on agricultural land. When feeding it maintains its tail in an elevated position. This cowbird is the commonest and most widespread of the three North American species and is abundant throughout Florida, where it is a migrant and winter visitor to the south.

Male (top); female (bottom)

Shiny Cowbird *Molothrus bonariensis* 18–20cm

Male

Female

This parasitic species, from South America, has recently invaded the southern USA, arriving in southern Florida, by way of the West Indies, in the mid-1980s. The black male has a purple gloss to the head, neck, breast and mantle, while the tail is of medium length and the eye is brown. The female is gray-brown above and brownish below. Gregarious, it freely associates with the larger grackles. It feeds on grain, seeds and insects. Lays its eggs in the nests of other species, especially grackles and orioles. Frequents all habitats.

Orchard Oriole *Icterus spurius* 15–18cm

Male

Female

Distinctively plumaged males are immediately recognizable, but females could be confused with several other female orioles. Adult males have a black head and upperparts, while the underparts and rump are burnt deep chestnut-orange. Females and immatures are olive-green above and yellow below, with two indistinct wingbars. This species swings its tail from side to side. Prefers open country with scattered trees. Arboreal, feeding on insects, fruit and nectar. An uncommon migrant through southern Florida usually encountered along the Gulf coast; it is also a rare wintering species.

Spot-breasted Oriole *Icterus pectoralis* 24cm

Bruce Hallett

Adults have a black face, throat, back and distinctive spots on the breast. The head and underparts are bright orange, as well as the rump, while immatures have a gray-green back and a black bib. It is an urban dweller, which feeds on nectar and insects, and is especially fond of avocado, citrus and mango blossoms. It is locally common in parks and gardens, and is the only oriole that occurs in southern Florida during summer. Introduced to our region in 1948, it has been an established common resident of the lower east coast since the mid-1970s.

Northern Oriole *Icterus galbula* 18–20cm

Male

Female

Formerly known as the Baltimore Oriole, the male has a black head, back and wings with white wingbars. The rest of the plumage is bright orange. The female is orange-yellow below and olive above with white wingbars. It frequents hammocks, coastal forests and gardens. Feeds on insects, fruit and nectar. The Bullock's form, which breeds in the western USA and is rare in our region, has an orange eyebrow, less black on the face and a large white wing patch; it has enjoyed a checkered taxonomic history, but is currently considered a separate species. Northern Oriole is an uncommon migrant and winter visitor to southern Florida.

American Goldfinch *Carduelis tristis* 11–14cm

Male

Female

Locally called the wild canary, the spring male is vivid yellow with a black cap and forehead; the wings are black with white wingbars and the tail is black. Females and winter plumaged birds have yellow-olive upperparts, with a yellow wash to the head, and the wings are black with reduced wingbars. Occurs in pines, brushy fields and urban areas. It travels in large flocks and is commonly seen at backyard feeders. Feeds on a variety of small seeds, but prefers thistle. The goldfinch is a common and widespread migrant and winter visitor to southern Florida. Its high-pitched flight notes are distinctive.

Glossary

arboreal	living in and about trees
brackish	slightly salty
canopy	treetops
coniferous	pertaining to evergreens especially pines
crepuscular	by twilight
crown	treetops
endemic	a species found nowhere else in the world
emergent vegetation	plants whose roots are submerged, and the foilage emerges out of the water
everglades	wet, flat sawgrass and cypress prairie of extreme southern Florida
flight feathers	the long feathers of the wing
frugiverous	feeding exclusively on fruit
glean	gather
hammock	island-like stands of palms, cedars, or oaks which are encompassed by pine flatlands, marshes or prairies
insectiverous	feeding exclusively on insects
key	small coastal islands also known as cays
lore	the area between bill and eye
mangroves	three species of salt tolerant trees which form extensive coastal forests
mask	a distinctively colored patch of feathers around the eye
migrant	a species which takes long and regular journeys between its breeding and non-breeding areas
nape	that point where the back of the head merges with the neck
omniverous	feeds on a variety of plant and animal material
pelagic	oceanic

plunge-diving	diving for food from a height, and becoming completely submerged in the process.
primaries	flight feathers of the outer wing joint
resident	present in the general area for the whole of the year
sallies	leaping or flying from a perch to feed
salt marsh	flooded coastal area with little wave action
scapulars	feathers which cover the shoulders
scrub	plant community characterized by stunted oaks and woody shrubs which grow in sandy soils
secondaries	flight feathers of the inner portion of the wing
supercilium	a stripe over the eye
swamp	a poorly drained, frequently flooded forested area
understorey	the forest below the canopy
undertail coverts	small feathers which cover the bases of the tail feathers beneath the tail
uppertail coverts	small feathers which cover the bases of the tail feathers above the tail
wing coverts	small feathers which cover the bases of the primaries and secondaries

Further reading

Flieg, G. and A. Sander, *A Photographic Guide to Birds of the West Indies*, Ralph Curtis Books, Sanibel FL, 2000 and New Holland Publishers (UK) Ltd, 2000

National Geographic Society, *Field Guide to the Birds of North America*, 3rd Edition, National Geographic Society, Washington DC, 1999

Pranty, B., *A Birder's Guide to South Florida*, American Birding Association, Colorado Springs CO, 1996

Raffaele, H. *et al.*, *A Guide to the Birds of the West Indies*, Princeton University Press, Princeton NJ, 1998

Toops, C. and W. Dilley, *Birds of South Florida*, River Road Press, Conway AR, 1986

Useful addresses

Florida Audubon Society
1331 Palmetto Avenue
Suite 110, Winter Park FL 32789
407/539-5700

Florida Ornithological Society
3675 1st Avenue NW
Naples FL 33964

Everglades National Park
40001 SR-9336
Homestead FL 33034
305/242-7700

Corkscrew Swamp Sanctuary
375 Sanctuary Road
Naples FL 33964
941/657-3771

Birding hot lines

Florida Birding Report	561/340-0079
Miami Area	305/567-7337
Lower Keys	305/294-3438

Index